TEACHING AND LEARNING N

D0512898

Related titles from Cassell:

Julia Anghileri (ed.): *Children's Mathematical Thinking in the Primary Years*

Edith Briggs and Kathleen Shaw: *Maths Alive!*

Harry Daniels and Julia Anghileri: *Secondary School Mathematics and Special Needs*

Anthony Orton: *Insights into Teaching Mathematics*

Anthony Orton: *Issues in Teaching Mathematics*

Anthony Orton: *Learning Mathematics*

Anthony Orton (ed.): *Pattern in the Teaching and Learning of Mathematics*

Joanna Swann and John Pratt (eds): *Improving Education: Method and Research*

Teaching and Learning Mathematics

A Teacher's Guide to Recent Research

Marilyn Nickson

CASSELL
London and New York

Cassell

Wellington House
125 Strand
London WC2R 0BB

370 Lexington Avenue
New York
NY 10017–6550

First published 2000

British Library Cataloguing-in-Publication Data
A catalogue record for this book is available from the British Library.

ISBN   0-304-70618-3 (hardback)
       0-304-70619-1 (paperback)

Typeset by Kenneth Burnley, Wirral, Cheshire.
Printed and bound in Great Britain by TJ International Press Ltd, Padstow, Cornwall

Contents

Preface

Research in mathematics education always has its roots in the classroom. The stimulus for the research and the intended applicability of the findings are focused towards the improvement of the teaching and learning of mathematics. However, the communities of school mathematics teachers and researchers have a fairly small intersection in membership and media for common reading. The research community is driven by different goals, including publishing in journals of the academic community and presenting their work at international conferences, and most are involved in teacher education. Teachers are driven by what is directly useful in helping them make sense of what happens in the classroom and what can help them develop better teaching and learning methods and materials for their students. But these communities are clearly not very far apart. In the UK and in many other countries, mathematics education researchers and teacher educators often have long careers as schoolteachers behind them. They share the task of the training of teachers and they generally share the same concerns about educational policy.

What is needed, therefore, is a range of opportunities for bringing the two communities together, so that teachers can find out about research findings that can be useful to them and researchers can collaborate with teachers in researching the issues that they feel need addressing in the classroom. There are conferences appropriate to both, such as: the quadrennial International Congress on Mathematical Education; the annual meetings of the National Council of Teachers of Mathematics in the USA; the biennial British Congress on Mathematical Education; the Association for Mathematics Education of South Africa, and similar groups in Portugal, Israel, Australia and many other countries. Student teachers during their training, teachers involved in research projects or taking higher degrees, provide other opportunities. This book is a welcome and important contribution both to bringing research findings to the attention of all of those groups and to fostering a dialogue between teachers and researchers and between teaching and researching.

A key element in research, too often tacit, is its theoretical orientation. What assumptions are made about the learning process? What is perceived to be the role of the teacher? How does instruction affect developing cognition? How does one determine the order of content? At different times there are particular orthodoxies in teaching and learning, such as teacher exposition, active learning, group work, or whole-class teaching. What are the effects of these various forms of classroom interaction? Who benefits and who loses in each form? Why do they appear? What are the relations between assessment and learning? Whilst it is generally accepted that effective learning has to begin from the child's intellectual situation, what can be called 'weak' constructivism, what is the nature of the significance of the child's situation? Social and cultural factors are increasingly featuring in teaching and research in mathematics. How are we to take account of these crucial elements in the make up of individuals in teaching and learning mathematics?

The recognition of the connection between theoretical frameworks and assumptions on the one hand, and the structure and findings of research on the other, has always been a characteristic of Dr Nickson's work. In particular, she has been at the forefront of bringing social and cultural factors into the agenda of the research community. She was co-founder, for instance, of the pioneering UK group for Research into Social Perspectives of Mathematics Education in 1986, producing a book of papers six years later. It is not easy, however, to carry off the task of bringing research findings to the notice of teachers, to indicate implications for teaching, and to provide the theoretical background to that work in a clear, coherent and substantive manner. Without those connections, reports of research findings are much weaker and less valuable. What is more, readers, both teachers and researchers, should take a critical stand in reading about research, and the theoretical background is essential to such a reading. Dr Nickson has achieved those almost impossible multiple aims admirably, given the short time and the limit on length of a book of this kind. She has chosen to rely heavily on proceedings of conferences, a risky approach given the danger of criticisms of lack of rigour ascribed to such sources, but I believe she has succeeded in bringing to our notice valuable work that can help to improve the teaching and learning of mathematics.

STEPHEN LERMAN
Centre for Mathematics Education
South Bank University

In memory of

Stieg Mellin-Olsen

colleague and friend

Introduction

Mathematics teachers at all levels would be the first to argue that their subject has a special significance in the lives of the young people they teach. It should follow that teachers are as well informed as possible about current developments in the teaching and learning of the subject. However, the need for accessibility of research findings for teachers and educators at all levels is as great as it ever was. In the UK for example, two reports relating to educational research have recently been commissioned, one by the Department for Education and Employment and the other by the Institute of Employment Studies. The fact that two different bodies should recognize that there is a question to be answered in relation to research about schools and teaching and learning, suggests that there is still a need for the results of studies to reach the practitioners and the policy-makers whom they are intended to inform. Findings apparently are not reported in a way that is accessible, and teachers in particular are not well served in their intention to remain as professionally aware as possible by being well informed. It is arguable that when mathematics is the focus for research, this failure is particularly crucial. Teachers in many countries now have the mathematics curriculum with specific content areas and levels to be attained by learners identified for them, so that for many the important questions are not about what is to be taught and when, but how it can be taught more effectively.

The aim of this book is to offer a review of some of the current research. While most research builds on the results of past work, the intention here is to attempt to present a selection of studies which reflects the changes in thought and practice that have developed in recent years. The studies come from a wide variety of countries – evidence of the establishment of a vibrant community of researchers in mathematics education; and the content of the studies, on the whole, indicates strong commonalities of concern. Most of the work reported here has been carried out by teacher-researchers together with research practitioners but seldom by an individual researcher. This, in itself, reflects the change of focus in mathematics education over the past decade.

Many of the studies have appeared as papers in the proceedings of international conferences which tend to be couched in terms that are sometimes more accessible than those appearing in journals. Studies of this kind are selected on the basis of quality and relevance by international committees of mathematics educators and reflect a rigour of a different nature from that of more formal papers that appear in research journals. They have to be made meaningful to a live audience, and this in itself may account for a greater degree of accessibility.

The themes of the chapters were chosen to incorporate as wide a spectrum as possible of topics found in the body of research, with the exception of probability and statistics. It was considered important to include the latter deliberately to highlight how little is known about the teaching and learning of the topic, in spite of the fact that it is seen as highly important in contemporary life and is one of the major areas for the application of number. The proscribed length of the book has meant that a whole chapter could not be devoted entirely to problem-solving, and this topic is discussed within the other themes as far as possible. The five chapters are Number and Applications, Shape and Space, Probability and Statistics, Algebra, and Classroom Research. Each chapter ends with the identification of some implications for teaching which are intended to flag possible points that have general relevance for teaching and clearly are not meant in any way to be prescriptive. While the studies reported all hold some important indicators for teaching and learning mathematics, there are three things to be remembered:

1. the selection of studies is inevitably a personal one;
2. the studies form a very small (but hopefully representative) sample of the full body of research in the field;
3. the implications drawn are affected by both of these factors.

Readers of this book will quickly become aware that researchers in mathematics education are concerned about the tensions and anxiety the subject raises in both teachers and learners, and the impact these have on the effectiveness of classroom teaching and learning. At the same time, mathematics educators are seeing the subject itself in a different light, with an increased emphasis on doing mathematics and recognizing when mathematics can help solve a problem, selecting the appropriate mathematics and then using it to reach a solution. Being put in a position of having to mathematize leads to meaningful learning. However, there is also the recognition of the need for tools to use in reaching a solution, and numeracy is one of the keys in this respect. There are facts to be learned in some cases where understanding follows later and the reason for knowing them becomes clear. Readers will also note that mathematics educators are aware of the importance of the pivotal role their subject plays in modern life; this emphasizes further how vital accessibility to mathematics is for *all* learners, a factor which increases the tension for both teachers and learners.

Finally, it is important to emphasize another kind of tension related to research in mathematics education: that between the research that informs policy and the research that informs practice, and whether they are always necessarily one and the same. There is the 'demand side' and the 'supply side' as noted in one of the reports about educational research mentioned earlier, where the demand side is seen as the teachers together with the policy-makers, and the supply side is seen as the researchers. It is conceivable that some research will be relevant to teachers that will have little to do with policy-making. On the other hand, there is no research that is relevant to policy-makers that is not relevant to teachers. Problems arise when there is a conflict between the two, and it is probable that with a well-informed teaching profession, such conflict is less likely to occur.

In writing about the influences brought to bear in pre-service and in-service teacher education, the Chief Inspector of Schools in the UK has said that 'The findings of educational research are important because for better or for worse they shape these influences and, in doing so, help to define the intellectual context within which all involved in education work' (Tooley and Darby 1999, p. 1). In order to influence the teaching and learning of mathematics more positively, a greater degree of transparency in research in mathematics education is needed that allows all participants to become more professionally informed. Educational researchers are accountable to the community which they serve through the accessibility of their work, and their contribution is measured according to the degree to which they achieve this. The intention of this book is to contribute to this accessibility.

Chapter 1

Number and Calculation

SHIFTS IN RESEARCH PERSPECTIVES

It is important when considering any research to have a background against which to interpret it. The theoretical perspective that informs the researchers' work is necessary in order to understand what they choose to study, and why and how they choose to study it. This chapter will begin with a brief overview of shifts in research perspectives in the teaching and learning of mathematics generally, before going on to consider research that concerns the teaching and learning of specific concepts related to number, the four operations and measures.

These theoretical perspectives arise from a change in emphasis in mathematics education that has taken place over the past decade and a half. For many years up until that time, the theoretical underpinning of much of the teaching and learning of mathematics had been heavily influenced by Piagetian thought, with some lingering influences from behaviourism, and as a result, much of the research tended to focus on the individual child learning specific concepts in an isolated situation. The shift away from the dominance of this approach has been brought about partly by the recognition that there were certain limitations in Piaget's work, one of these being that the experimental work that informed his theory did not take the social and cultural factors of the learning context into account. There have also been research findings that have challenged specific factors related to mathematical learning, such as the stage at which children are able to count and to symbolize (Hughes 1986). This is not to say that Piaget's work has been discounted, but rather that it has been tried, evaluated, adapted and has provided the basis for new theories in the way in which theories grow (Popper 1972). These in turn have provided the springboards for many of the newer approaches used in mathematics education, as we shall see.

The most fundamental change which Piaget has brought to education with his concept of a genetic epistemology is identified by von Glaserfeld (1991) as a redefinition of the concept of knowledge. This redefinition has influenced mathematics education in particular, and has redirected us towards a more pragmatic view of how knowledge comes into being, bringing with it an adaptive function which has the following effect:

> This means that the results of our cognitive efforts have the purpose of helping us to cope in the world of our experience, rather than the traditional goal of furnishing an 'objective' representation of a world as it might 'exist' apart from us and our experience. (von Glaserfeld, p. xv)

This reminds us that not only is learning a purposeful activity, but that if we want children to learn in a meaningful way, the purpose must be relevant to their growth and adaptation to the world around them. Thus the somewhat mechanistic approach of behaviourism has been overtaken by one with an emphasis on cognition and learning as 'coming to know' as opposed to 'coming to do' certain things.

Concurrently with this shift from the psychologically-oriented view to a more socio-psychological interpretation of how learning takes place, there has been a similar shift in philosophical issues related to mathematics education. This has involved a change in perceptions of the nature of mathematics from a 'given' abstract body of knowledge to one that is seen as social in its origins and its applications (Lerman 1990, Ernest 1991, Restivo 1991, Nickson 1992, Nickson and Lerman 1992). It is recognized that mathematics, just as any other subject, has its origins in human activity, and, as a subject, it grows and changes as a result of problem-solving, trial and error and the interpersonal exchange of ideas. As with other subjects, it is just as possible for some part of mathematics either to be disproved and discarded or for some new mathematical idea to be discovered (e.g. the recent solution to Fermat's last theorem), so that the mathematical body of knowledge is not static and inert but changing and expanding. At the same time, mathematics education researchers have become more aware of the difference in demands made by mathematics as a way of thinking, and education as a discipline (Brown 1997) and of the need to accommodate the two.

These two phenomena taken together have led research in mathematics education to move away from an emphasis on experimental situations where children's mathematical thinking is studied in an artificial situation with controlled variables, to one where children's mathematical thinking is more likely to be studied in classroom or group situations and the process of learning mathematics becomes one of discussion, negotiation and shared meanings. In this way, the aim is not just for the pedagogy to be effective but it will also convey messages about the nature of mathematics itself, its relevance to everyday life and its foundation in human activity. The intention is that the subject comes to be perceived as social in its foundations as well as its uses (Nickson and Lerman 1992). Conjecture, hypothesis, testing ideas, trial and error all become part of the taken-for-granted aspects of mathematics. The focus on the child is retained but it is on the child-in-context, i.e. the child as learner who has had considerable experience in the context of the social world before entering the classroom, experience which is continuous and contiguous with their experience at school, and which will affect what and how they learn in the context of the classroom. In a sense, children bring their own individual context with them.

New directions in research in mathematics education have resulted, overall, in increased attention being paid to socio-cultural factors. At the broader level, the society in which a given mathematics classroom is situated affects the teaching and learning that takes place, as well as the mathematics that is taught (Scott-Hodgetts 1992). At the level of the individual, pupils have their particular view of what mathematics is all about (Winter 1992), and the cultural diversity within a class may mean that some of them bring a different first language to the learning situation (Clarkson and Dawe 1997) as well as different approaches to understanding concepts such as number, that are culturally determined (Ginsburg *et al.* 1997). At the same time, teachers struggle with their own view of mathematics and how to get it across in the classroom in such a way as not to conflict with the pupils' world (Adler 1997). This world is essentially the community in which they live and forms much of their reality, and mathematics will be meaningful (or not) if children see its relevance to their world (Skovsmose 1994). Accordingly, mathematics has to be 'realistic' to be accessible to them (Gravemeijer 1997). It is possible for at least a large majority of children to gain access to acceptable levels of mathematics and to be properly numerate (see below) only if factors of this kind are paid some attention. This will become evident in much of the research that is reported below.

NEW APPROACHES AND THE TEACHING OF NUMBER

The approaches to teaching number (as well as other areas of mathematics) that have been developed in order to address these concerns are founded in cognitivism, which is essentially concerned with 'how we come to know'. Examples of some of these approaches are briefly described below, together with examples of studies which illustrate their influence on aspects of the teaching and learning of number. The notion of *constructivism* is pervasive since each has evolved from the basic Piagetian notion that individuals *construct* their knowledge as they interact with the reality of their world. (It is important also, however, to recognize the work of sociologists of knowledge such as Berger and Luckman (1966) and their notion of the 'social construction of reality'.) The subtle differences between constructivist approaches lie in the perspective of personal *construction* of mathematical knowledge, or the personal *reconstruction* of mathematics that already exists. The former is sometimes referred to as the 'bottom up' approach while the latter is known as 'top down' (Gravemeijer 1997). We shall now look at some of these approaches and the different emphases each of them has, and, in considering such theories, bear in mind 'the reality of the pedagogic situation in which each of us works' (Nickson 1995).

Constructivism

A large proportion of current research literature in mathematics education is given over to discussions of constructivism and studies related to it. The term has taken on a global meaning which, with frequent and uncritical use, could very easily have negative effects in the classroom if the underlying ideas are over-simplified in their use. (In an attempt to redress this potential it is frequently referred to as socio-contructivism or social constructivism, e.g. Cobb *et al.* 1991, Ernest 1991.) As some of the research reported here will indicate, 'this new model of learning has yet to be situated properly in the totality of mathematics education' (Nickson 1995, p. 165). Noddings refers to it as 'a movement growing in influence and popularity' and warns that it may come to be seen as a 'method' as opposed to a theoretical perspective (Noddings 1993, p. 158). It is important to bear these considerations in mind when observing the ways in which constructivism exerts an influence in the mathematics classroom. In referring to constructivism as an epistemology, Cobb *et al.* (1991) note that:

> Its value to mathematics education will, in the long run, depend on whether this way of sense making, of problem posing and solving, contributes to the improvement of mathematics teaching and learning in typical classrooms with characteristic teachers. If it eventually fails to do so, then it will become irrelevant to mathematics educators. (Cobb, Wood and Yackel 1991, p. 158)

A central theme of the constructivist approach is an acceptance of the fact that the reality of one individual is different from that of another and that individuals 'construct their own mental representations of situations, events, and conceptual structures' (*ibid.*). Children clearly already have such representations when they begin schooling, and recognition of this fact is paramount in the constructivist perspective of what should go on in the classroom. What becomes accepted generally as 'knowledge' is the result of shared consensus over the years: as, for example, our number system has become. What is 'knowledge' for an individual child is what they have experienced and stored and made their own. They extend that knowledge as a result of new experiences when they have to use what they already know to interpret the new situation in which they find themselves. As a result, the social aspect of learning becomes vital to the development of knowledge (Jaworski 1994). The idea of the child constructing their own knowledge does not convey an impression that promotes the social aspect of learning. However, Cobb *et al.* (1991) write that:

> Because mathematical meaning is inherently dependent on the construction of consensual domains, the activities of teaching and learning must necessarily be guided by obligations that are created and regenerated through social interaction. (Cobb, Wood and Yackel 1991, p. 165)

This may go some way towards acknowledging the social factors inherent in mathematical knowledge-making, but for some it is not far enough. Lerman (1998) writes:

A psychology focused on the individual making her/his own sense of
the world does not engage with social and cultural life: other theoretical
discourses, such as approaches to sociology which merely describe are
not adequate for mathematics education either. (Lerman 1998, p. 77)

Lerman advocates what he calls a 'discursive psychology for mathematics
education' which he envisages incorporating 'action, goals, affect, power and its
lack, based on sociocultural origin' and is discussed further in Chapter 5 (*ibid.*).

It is important to recognize these variations of the constructivist
perspective and to be aware of the effect of the different interpretations of the
way in which mathematics is taught and learned. There are three major
implications of a constructivist approach for the effective teaching and
learning of mathematics. The first is that children are no longer seen as
receivers of knowledge but *makers* of it, in that they are actively engaged in
selecting, absorbing and adjusting what they experience in the world around
them. It follows, then, that to learn mathematics, children must be put in
situations where they have to mathematize and so be involved in doing it. The
second major implication of constructivism follows from this which is that, in
order to mathematize, children need to experience mathematics in a context
other than a purely mathematical one. In order to make sense of the
mathematics they meet in school, to access it and make it their own, they have
to link it with the reality of their world and what they already know (Aubrey
1993). The teaching of mathematics must therefore be linked with contexts that
are 'real' to children and with which they can identify. The third major
implication is that each child's contribution in a mathematics lesson needs to
be acknowledged and considered, not just by the teacher but by other members
of the class. This follows from the unique nature of each child's perception of a
situation and the experience they bring to it. Aubrey suggests that every child
must be helped to extend their range of mathematical strategies, and
mathematical teaching should not undermine 'their natural inventiveness by a
struggle to find the single, convergent and acceptable response' (p. 39). This
encapsulates one of the fundamental aspects of what constructivism is about.

Gravemeijer (1997) points out that, 'Educationalists attach different
consequences to the recognition of the importance of informal strategies and
situated cognition' (p. 319). As a result of this importance attached to the
individual's perceptions of mathematics it is held that:

mathematics education should acknowledge idiosyncratic constructs
and foster a classroom atmosphere where mathematical meaning,
interpretation, and procedures are explicitly negotiated. (Gravemeijer
1997, p. 320)

This is reflected in the extension of the concept of constructivism to either
'socio-constructivism' or 'social constructivism' noted earlier (see p. 3).

The children's strategies and solutions in solving problems are the focus of
the teaching. They are shared and discussed and, in the process, become

justified and accepted or not, as the case may be, and the meaning develops in the process of this reflection and discussion. Much of the research that is touched upon later in this chapter reflects this perspective.

Cognitive Guided Instruction

Cognitive Guided Instruction (CGI) (Carpenter *et al.* 1989) is a project in which a particular approach is embedded that focuses on the identification of the strategies children use as opposed to an interpretation of how children learn. It has its basis in a cognitive perspective which takes on board the importance of children's informal strategies in constructing their own knowledge, and the intended outcome is to provide teachers with support in building on these strategies. The following four fundamental assumptions underlying CGI have been extracted from the literature by Behr *et al.* (1992):

1. Children construct their own mathematical knowledge.
2. Mathematics instruction should be organized to facilitate children's construction of knowledge.
3. Children's development of mathematical ideas should provide the basis for sequencing topics of instruction.
4. Mathematical skills should be taught in relation to understanding and problem-solving. (Behr *et al.* 1992, p. 325)

Research is carried out to identify the strategies children use in the classroom situation and this is made available to teachers to provide them with a framework for their teaching. They can draw on these strategies to use in their teaching and they will also presumably be able to recognize more easily what is happening in their mathematics classrooms and to use children's idiosyncratic methods to better advantage in helping them to learn. A study of a classroom in which this approach was used appears below (see Steinberg 1994).

Cognitive apprenticeship

Another approach within the cognitivist paradigm is the 'cognitive apprenticeship' model (Brown, Collins and Duguid 1989) where knowledge is seen as 'situated, being in part a product of the activity, the context, and culture in which it is developed and used' (p. 32). The notion of situated learning developed by Brown *et al.* (1989) and Lave and Wenger (1991), plays a strong part in current research in mathematics education generally (e.g. Boaler 1993, Nunes *et al.* 1993). A major implication of this perspective for learning and teaching mathematics is that to be meaningful, new mathematical knowledge and skills are most effectively learned in situations where they are applied. In their discussion of situated learning, Fennema and Franke (1992) note that 'Knowledge acquired in school is not situated in the broader life of an individual becasue the activities, contexts, and culture of the school are not related by the learner to his or her out-of-school culture' (p. 160). The teacher's role becomes one of helping children to make connections between the learning that takes place in the classroom and their experience in the 'real world'. In a somewhat blunt comparison, the difference between this approach and constructivism would be that it is seen as a 'top down' perspective where the mathematics is taken as 'given' and children do not actually construct it themselves. However, there is commonality between the two in the importance given to making use of the children's prior knowledge and experience and to the teacher's role of acting as a mediator of that experience so that mathematics is learned in a meaningful way.

Realistic mathematics education

A fourth approach is known as 'realistic mathematics education' which has developed from Freudenthal's view of mathematics as a form of activity and of mathematics education as a process of 'guided reinvention' (Freudenthal 1973). 'Mathematizing' is seen as the goal of mathematics education and involves solving problems, looking for problems and organizing subject matter. Mathematizing may be done within mathematics or within reality; the problems may be non-contextualized mathematics or they may be everyday problems such as finding the cost of a number of items at a given price. Realistic mathematics is mainly concerned with generalizing (analogues, classifying and structuring) and formalizing (modelling, symbolizing, schematizing and defining) (Gravemeijer 1997).

Five educational principles implicit in realistic mathematics education have been identified by Streefland (1991b) in the context of teaching and learning fractions and provide an example of the basis of this approach:

1. the source of concept formation is reality;
2. pupils are given the opportunity to be constructors and actively contribute to the learning process by this involvement;
3. the learning process must be interactive so that in the course of constructing knowledge about fractions from real life situations, pupils discuss and collaborate when necessary;
4. different lines of learning (e.g. fractions, ratio and proportion) are entwined so that both vertical and horizontal mathematization can take place;
5. the various tools used in the process of coming to understand mathematics symbols, diagrams and visual models, should result from the need to describe and use what they have found out for themselves.

Clearly, this has much in common with other constructivist approaches but, as noted earlier, it has a 'top down' connotation in that it acknowledges the 'givenness' of mathematics at the outset while, at the same time, it aims to build on children's strategies and previous learning to gain access to it. However, implicit in this approach is an acknowledgement of 'the importance of *knowing* number facts' and that, in itself, emphasis on mental strategies does not necessarily strengthen basic number skills (Beishuizen 1997, p. 18).

Critical mathematics education

The increase in the development of social considerations related to mathematics education noted earlier is exemplified by Skovsmose's (1994) work in developing a philosophy of what he calls 'critical mathematics education'. This philosophy has implications for the nature of the mathematics curriculum, in terms of aims, content and pedagogy which can be seen in much of current mathematics education thinking (e.g, Restivo 1993, Noddings 1993, Gates 1997, Lerman 1998). It is based on the notion arising from developments in critical theory and summarizes the relationship of critical theory to education in this way: *'If educational practice and research are to be critical, they must address conflicts and crises in society'* (Skovsmose, p. 22, author's italics). He sees mathematics education as playing a strong role in developing 'critical citizenship' so that it does not 'degenerate into a way of effectively socializing students into a technological society and at the same time annihilating the possibility of them developing a critical attitude towards precisely this society' (p. 59).

The examples given by Skovsmose of the implications for the mathematics curriculum of this theory are not unlike the approach used in realistic mathematics education and are in the form of project work. However, the importance of the reflective aspect of what is entailed is emphasized to a greater extent, both in depth and breadth of scope. Pupils are not only asked to reflect upon the mathematics inherent in the project, they are invited to reflect upon:

(a) the *social issues* implicit in the situation which forms the context of the project;
(b) how the mathematics they have engaged in has helped to inform the *judgements* they have made; and
(c) the *consequences* of certain actions being taken as a result of the judgements they have made.

Skovsmose refers to the use of a 'thematic approach' to the mathematics curriculum and he describes some of the work that has been undertaken in Danish high schools that exemplify this (see also Nielsen *et al.* 1999). One example is a project dealing with 'Economic Relationships in the World of a Child' which is made up of units dealing with the immediate world of the individual child (starting with the pocket money they get each week), then the child as part of a family, and finally as a member of society.

While the notion of critical mathematics education may be new to many, there has been a growing acknowledgement of the power that mathematical knowledge brings to the individual in a technological society, and how important it is for all mathematics educators to be aware of this, whatever the age of the children they teach (e.g. Mellin-Olsen 1987, Knijnik 1995, Julie 1998). Noddings (1993) hints at this when writing about constructivism; she says 'Unless it is embedded in an encompassing, moral position on education, it risks categorization as a *method*, as something that will produce enhanced traditional results' (Noddings 1993, p. 158, author's italics). In other words, the *status quo* will be maintained; the mathematics educators' aim of educating children in mathematics not just more effectively in the sense of being numerate, but more meaningfully in the sense of being aware of the part it plays in everyday decision-making, will not happen. Skovsmose's work provides a theoretical perspective from which this can happen.

NUMERACY

As well as being aware of theoretical perspectives in interpreting recent research in number in particular, it is also important to be aware of current interpretations of numeracy. One definition of basic numeracy is given as 'the ability to operate flexibly with numbers in solving real problems, and especially in operating efficiently either mentally or using a calculator' (Dickson *et al.* 1984, p. 250). A broader and more recent working definition for a numeracy project states that 'Numeracy is the ability to process, communicate and interpret numerical information in a variety of contexts' (Askew *et al.* 1997, p. 25). More recently still, the document issued to Primary schools in the UK that underpins the Numeracy Hour to be introduced in all schools says the following:

> Numeracy in Key Stage 1 and Key Stage 2 is a proficiency which involves confidence and competence with numbers and measures. It requires an understanding of the number system, a repertoire of computational skills and an inclination and ability to solve number problems in a variety of contexts. Numeracy also demands practical understanding of the ways in which information is gathered by counting and measuring, and is presented in graphs, diagrams, charts and tables. (Department for Education and the Environment 1999, p. 4)

There is a shift here from an ability merely to solve numerical problems to acknowledging the contextual implications inherent in numeracy. Communication and interpretation of information as well as 'practical understanding' (which suggests application) imply a social context that has not received the attention it was due in the past. This is implicit in the cultural model for numeracy advocated by Baker (1995) who argues that this is an appropriate perspective from which to view numeracy because mathematics is both the product of, and the basis for, the science, commerce and technology of a society. This shift in perspectives on numeracy is further evidence of the acknowledgement of the social concerns embedded in mathematics education and, in this case, at the most basic level of its learning and use.

NUMBER AND COUNTING SKILLS

The studies by Hughes (1986) following on work by Donaldson (1978) were foremost in the UK in challenging the results of some of Piaget's work, particularly with respect to children's learning of number skills. Hughes

(1986) found that children's 'counting strategies are frequently untaught, and are meaningful attempts by the child to solve the problems confronting them' (p. 35). He also found that 'even pre-school children are able to represent small quantities, either spontaneously, or with small amounts of prompting' and it should not be presumed that they have necessarily been trained by parents in meaningless parroting (p. 77). Although forms of representation mainly take the form of one-to-one correspondence and are pictographic or iconic, they are still abstractions of a physical counting situation.

Gray *et al.* (1997) use the activity and representation of counting in their development of a cognitive theory related to children's conceptual development in mathematics, and devised the notion of *procept* to take into account the role of mathematical symbolization where it can represent a process (to do something) or a concept (to know something). They approach this through a consideration of how children learn to count and to use the names of numbers (see below). They note that learning to count is grounded in actions related to physical objects, but the physical objects themselves essentially need to be ignored: it is the *actions* carried out by the child that are important and which have to create an 'object of the mind' (p. 115):

> For some there may be a cognitive shift from concrete to abstract in which the concept of number becomes conceived as a construct that can be manipulated in the mind. For others, however, meaning remains at an enactive level; elementary arithmetic remains a matter of performing or representing an action. (*idem*)

They set out to evolve a theory to explain what it is that children do differently in each case, and why, and central to this theory is the notion of the *proceptual divide*. In the process of learning to 'count-all', children first of all have to 'count-on'. It is at this stage of counting on that the proceptual divide occurs. Children who proceed to 'proceptual thinking' have a meaningful knowledge of number facts and can use counting procedures flexibly. Children who exhibit 'procedural thinking', use counting procedures and number facts which lack flexibility, and appear to have been learned in a rote manner and not in a meaningful way. Figure 1.1 is a diagrammatic representation of the proceptual divide.

Gray *et al.* (1997) postulate that although this divide may occur as a result of pedagogy, there may be more subtle reasons that have to do with a more qualitative aspect of children's thinking which 'on the one hand places the emphasis upon concrete objects and actions upon these objects, and on the other on abstraction and flexibility intrinsic within the encapsulated object' (p. 121). They focus on imagery, in the first instance, to try to explain this in work carried out with children from four age groups within the 8- to 12-year-old age range. The tasks given to the children involved responding to auditory and visual stimuli (the latter including symbolic representations such as '5' and '3 ÷ 4'), when they were asked to give a first response of what came to mind, and then were given about 30 seconds to talk about each. In their conclusions they

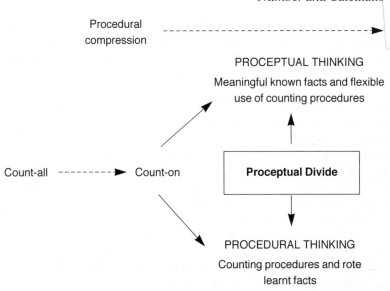

Figure 1.1: Procedural compression associated with numerical development and the formation of a proceptual divide for early number. (Gray et al. 1997, p. 121)

note that the low achievers tended to concretize and focus on all the information in a given situation, and state that 'Imagery in the numerical context is strongly associated with procedural aspects of numerical processes. The children carry out procedures in the mind as if they were carrying out procedures with perceptual items on the desk in front of them.' In contrast to this, the high achievers seemed to focus on abstractions which helped them to make choices and to reject information. The researchers conclude that:

> The ability to filter out information and see the strength of such a simple device as a mathematical symbol appears to be confined to the high achievers. The evidence suggests that children who are 'low achievers' in mathematics appear unable to detach themselves from the search for substance and meaning – no information is rejected, no surface feature filtered out. (p. 128)

They suggest that the problems low achievers have is not to do with a deficit in working memory, but rather 'Their problem is associated with its use and not its capacity' (p. 127).

Aubrey's (1993) work with a class of sixteen infants in an urban school in the UK supports this. She showed that all the children came to school with a substantial knowledge of number concepts and mathematical skills. The more able in the group were able, with a fair degree of success, to count and order objects up to 10, to read and write numbers up to 10 and to recognize and repeat pattern. At the very least, all children showed a 'range of informal strategies' even if they did not have particularly stable conventional mathematical knowledge (Aubrey 1993, p. 39). She makes the point that much of this prior knowledge children already have is what they are then 'taught' on entry to school, while what should be happening is that the strategies they already have should be extended. Her findings suggest that children's 'natural inventiveness' should not be undermined in a struggle to find a single, acceptable response. Mboyiya (1998) also stresses the need to recognize children's informal strategies, but emphasizes the importance of these strategies becoming refined.

Although children do have this prior knowledge, research into children's counting ability shows how difficult it can be for the teacher to judge the extent of a child's understanding of, and ability to use, number skills and then to decide how to build on this knowledge. To add to the complexity of the situation, their experience is variable both within and across tasks (Gelman and Meck 1986). An overview of research into what is involved when considering children's counting strategies alone is provided by Fuson (1992). She identifies the following five points children must learn. In developing their counting ability, children in each culture must:

1. learn the number sequence of their own culture;
2. learn the indicating act of their own culture (usually pointing);
3. learn to use the indicating act to connect one number label to one entity (make local correspondences);
4. learn methods to remember the already counted entities so that entities are not counted (make a global correspondence); and
5. learn the cardinal significance of counting. (p. 248)

The subtle differences between cultures in the way in which early number concepts are established was explored by Ginsburg *et al.* (1997) over a two-year period. They studied the differences in preschool mathematical knowledge of 476 4-year-old children in five countries (Japan, China, Colombia, Korea and the US) and among different ethnic groups in the US. A set of ten tasks based on a birthday party theme was given to groups of children who were matched in terms of socio-economic status and gender. The tasks included the following: recognizing number from finger displays; the perception of 'more'; production of numbers from 1 to 10; counting by ones; comparison of number; concrete addition; digit span memory; concrete subtraction; informal subtraction; and finally, informal addition. Their conclusion after comparing countries on total scores, was that 'Asians exhibit a better overall performance in informal mathematical thinking than do US children, and that US children perform at a higher level than do Colombian children' (p. 196). However, they emphasize that the differences between children from the US and the Asian countries is not large. They also note the following:

It is interesting but puzzling to note the tremendous variation in the strategies exhibited by the various groups. For example, in the

Concrete Addition task, Chinese children touch objects whereas Japanese children produce the answer quickly, without overt counting. (p. 197)

Among the US children, different ethnic groups used a variety of approaches, but where there was more consistent use of strategies, there was greater success, which seemed to reflect the situation in the Asian countries. They interpret this to suggest that consistency of use implies a strategy has been well learned and as a result, it can be used correctly. They found across the ethnic groups in the US sample (which included African Americans, Hispanics, Koreans and Whites) that children 'with the possible exception of the Hispanics, possess reasonable competence in informal mathematical thinking' (p. 191). They pose the question as to why some children go on to do well and others do not, and they note particularly the case of the African-Americans whose preschool performance is not very different from other groups including the Japanese. They begin school with an 'adequate intellectual ability but do not go on to realize their potential in school' (p. 201). Ginsburg *et al.* (1997) suggest that:

The educational system seems to amplify, rather than modulate, differences with which children enter school. Subordinate groups are generally provided with inferior education, including teachers who expect them to fail.

The final conclusion drawn is that 'In general, informal mathematical ability does not appear to be a major mediator of differences in school mathematics achievement'.

This consideration of the influence of culture is a reminder of the situatedness of all learning (D'Ambrosio 1985, Lave 1988, Brown *et al.* 1989) and the enormous task the teacher has in taking this into account in the everyday classroom situation (Nickson 1992). However, while there is ample evidence (e.g. Ginsburg *et al.* 1997, Aubrey 1993) that children come to school with a fair amount of informal knowledge related to number concepts, a note of caution is offered by Bryant (1997) who states that:

They make a definite start before they go to school. They know some things but definitely not everything by the time that they go to school. We must find out what effect this knowledge has on them when they enter the classroom. (p. 67)

Cardinality

The meaningfulness of counting depends upon an understanding of the cardinal aspect of number which Fuson (1992) mentions as the fifth and final point in the process of learning to count (see above). The ordinal aspect of number where counting is the process of ordering a sequence (first, second, third etc.) appears a simple matter compared with coming to an understanding of cardinality where a given number denotes a quantity. Ordinality answers the question 'Which one?' Cardinality answers the question 'How many?' Although they may seem to be equally simple questions, a point repeatedly established in the research is the difficulty some children have in progressing from an ordinal to a cardinal understanding of number, which can be a considerable obstacle in their mathematical development. There is an extensive research literature connected with how children learn to count and, in particular, how they come to understand the concept of cardinality (e.g. Gelman and Meck 1986, Carpenter 1986, Fuson 1992, Gray *et al.* 1995).

The difficulty in establishing an understanding of cardinality lies in learning that the word 'three' for example is not just a name but has a further meaning. Children have to move from a stage where they use number as 'word names' in a sequence, to using them in one-to-one correspondence with a set of objects (not necessarily arranged in a line where they count from left to right) and ultimately to using only one number word to give a name to a set with more than one object. This is a sophisticated undertaking (see Fuson above) but one which, as we have seen (Hughes 1986), very young children are capable of achieving. However, it is easy to think that children understand the cardinality of number when they in fact do not, a situation which Gelman and Meck (1986) describe as procedural competence masking conceptual competence. Without such a well-established understanding, children will not be able to carry out simple addition and subtraction in a meaningful way nor have a strong procedural basis for further mathematical development.

Gelman and Meck (1986) identify three types of competence children must have in learning to count:

1. conceptual;
2. procedural;
3. utilization.

They argue that it is the mastery and interplay of these three competences that help children to establish sound understanding of counting and ultimately the idea of cardinality. They suggest that 'even if young children have complete conceptual competence, there is no guarantee they will display it in their behaviour' (p. 40). That is to say, they may be able to count and appear to know something about the 'three-ness' of a set of three objects; this may hold in one situation where, for example, all three objects are apples, but not in another where the three objects are an apple, a banana and an orange. A child may be able to count the apple, banana and orange as '1, 2, 3' and conclude that there are three pieces of fruit in the set. But given an orange, apple and banana in that order, since banana was '2' the first time around, they may wrongly say there are two pieces of fruit in the set. This indicates both that their 'counting-as-naming' skills and grasp of the conservation of number have not gone hand-in-hand, but it also shows that their competence in counting is not 'order-indifferent'. As Baroody and Ginsburg (1986) put it, 'to discover that differently ordered counts of a set do not affect the outcome, children must be willing first to count sets in different orders' (p. 77). Until they do so, children will not develop the principles which guide their conceptual competency further to establish a procedural competency.

It is important to understand what is meant by the word 'principles' in this context. Gelman and Meck (1986) note evidence that suggests very young children 'monitor, rehearse and self-correct their performances' (p. 39). They can only 'self-correct' their actions if they have something stored in their minds by which to judge them. It is argued that these aspects of children's experience which must be represented in some way are stored in their subconscious, and it is these that form the principles that guide their actions. These principles contribute to the development of schema or schemata in children's learning which are constantly brought into play and expanded as they continue to learn.

For children to reach an understanding of cardinality, they must be given adequate experience of using their knowledge of 'numbers-as-names' in counting, i.e. their conceptual competence, in procedural ways. 'Competent procedures, that is, ones that are tied to conceptual competence, generate opportunities to learn principles the child does not know' (Gelman and Meck 1986, p. 54). With sound procedural competence linked with the conceptual, children's counting skills develop further to establish utilization competences, i.e. they bring the first two kinds of competence together and use them in identifying quantity in numerical operations. In other words, they begin with conceptual competence, move on to procedural, and eventually arrive at utilization competence when they have achieved an understanding of cardinality.

As indicated earlier, one of the difficulties some children may have in establishing the connections between the competences that lead to a grasp of cardinality can arise when, in the process of learning to answer the question 'How many?', children confuse the ordinal and cardinal aspects of number. When faced with such a question, children are invariably told by teachers to 'Count them.' Gelman and Meck (1986) believe that this act of 'tagging', as they call it, becomes confused with the outcome of the counting action and they suggest that children should be encouraged 'to keep separate their tagging and cardinal response goals' (p. 48). If they are encouraged to count (procedural) *and* are asked 'How many?' (utilization) at the same time, they may easily become confused and the situation in itself may mask a competence which they already have. One simple way to help overcome this confusion is to ask children to pause after their counting before they give the answer to 'How many?' This allows counting, which is a sub-goal of the whole procedure, to become integrated appropriately in the total action of arriving at an answer and to give the correct cardinal response.

To suggest that there is consensus in all the research about counting skills would be misleading, particularly where conceptual and procedural knowledge are concerned. Baroody and Ginsburg (1986) for example, do not consider conceptual knowledge necessary for the establishment of procedural knowledge. Carpenter (1986) helps us out of this corner by suggesting that 'it may not be productive to argue whether particular conceptual knowledge exists at a given point in time' (p. 121). What we do know is that conceptual knowledge of one kind (such as the names of numbers) is necessary for the acquisition of new conceptual knowledge (such as cardinality).

Teaching counting skills

Establishing mental counting techniques in children of 5 and 6 years of age was one of the focal points of a study reported by Womack and Williams (1998). An aim of the study was to familiarize the children with three aspects of counting: counting-on, counting-back and counting-up (finding how many numbers lie between two given numbers).

The study built upon the idea of children constructing counting skills in relation to their real environment which consisted of a stepping-stone number trail and provides an example of children being actively engaged in building up a concept. (Stepping stones in a stream were very much part of the geographical location of these children who lived in the Lake District in the UK). The only idea inherent in the counting situation was 'succession' or order, one number following another, and no extraneous ideas such as equidistance were introduced nor any other distracting mathematical concepts. The activities the children engaged in moved from the physical stepping activity to mental counting, and means of representation moved from oral to written words, and children's written signs to signs devised by the researchers. The signs were transformational and required some action to be taken to determine a quantity (in this case a number of steps) and ultimately took the form of an arrow pointing up (for forward) or down (for back). In the first instance, the children enacted the counting by walking on stones, 'stepping on' or 'stepping back' a given number of paces. The next stage took the children indoors where stones were represented on a table-top and this time were numbered. Children moved a coin or other object to replicate the stepping action, from one stone to another. Further stages were developed which ultimately provided the children with a meaningful way in which to develop mental counting strategies. These, in turn, led to simple computation by using the arithmetic of transformational problems (where number operations are involved), and allowed children to use their own signs for the different operations. The whole process was built on using 'the children's intuitive transformational arithmetic' in order to prepare them 'for a firm and confident understanding of the conventional system of arithmetic' (Womack and Williams 1998, p. 185).

The benefits of an approach of this kind is that children are being led, through counting activities, to an understanding of simple arithmetic transformations initially through their own actions and in a familiar context. They then translate their active experience through different stages of abstraction which involve them in discussion and sharing ideas and coming to an agreement about the kind of symbols to be used. The symbols become associated with something the children have acted out and they are able to connect the abstract representation directly with their experience and give meaning to it.

Writing numerals

One difficulty children have in learning numerals is 'number names' or what Hewitt and Brown (1998) call 'the language of number words' (p. 41). They worked with pupils aged 11 to 14 years who were having problems with mathematics that arose from their lack of confidence with basic number. They give an example of the nature of the difficulty when one of the pupils is described as solving 67 – 41 by counting backwards. Rather than approaching such problems with number by reverting to concrete materials and attacking place value, they decided to focus on building pupils' confidence in 'being able to say and write whole numbers, and develop mental strategies for adding and subtracting whole numbers which encourage alternative approaches to those of count-on and count-back' (p. 42). Drawing on the work of Wigley (1997) they note the irregularity of the language used in counting past ten where, for example, the word 'thirteen' does not indicate that it is made up of one 10 and one 3. They decided to use a Chinese version of English word names such as that discussed by Fuson and Kwon (1991) where the name indicates the structure of the number, and for example, thirteen become 'one-ty three' to reflect how the number is made up. In other words, '-ty' becomes 'ten' throughout the whole number naming system. The reported study describes the growing awareness on the part of the teachers and researchers in the study of the extent to which this aspect of mathe-matical language plays a part in the learning difficulties some children have. They realize that:

> While the students still need to develop their knowledge in areas that cannot be worked on through the language, such as number bonds and partitioning, our work suggests that a lack of such number skills does not have to be a barrier to developing flexible skills for mental addition of larger numbers. (p. 44)

Wright (1998) also probed this aspect of mathematical learning when he set out to find what strategies children use to identify, recognize and write numerals, and the difficulties they have in acquiring these skills. The work was carried out with 5- to 7-year-olds and provides evidence of the complexities involved in achieving a grasp of the cardinality of number. Individual children were videotaped in teaching and assessment situations and the observations made from the analysis of the taped sessions were categorized. The kinds of difficulties children had included:

- being able to identify numerals when presented in a sequence but not out of sequence;
- saying the number word sequence from 1 to 10 but at the same time being able to identify the numerals in order when presented with them in random order;
- writing the right-hand digit first in numerals 13 to 19 (but writing the left-hand digit first thereafter);
- using digit names to identify two-digit numerals, e.g. twenty-eight for 28 (a system which breaks down for numbers such as 12);
- identifying a two-digit numeral in the reverse order, e.g. 27 as seventy-two;
- given any numeral, identifying it by counting forward, e.g. for 8 saying 'one, two ... eight';
- being able to write a numeral only after counting forward, e.g. 'one, two, ... seven' and then writing the numeral 7;
- difficulty identifying orally numerals from 1 to 10 (unable to generate the sound image that corresponded with the numeral);
- specific difficulties associated with numeral 12;
- in solving arithmetical tasks, thinking in terms of visualized numerals when they might be expected to be using number words.

The variety of difficulties displayed by the children in this study help to pinpoint the kinds of informal strategies that children use but also the misconceptions that may be inherent in some of these. While it is important to build on these strategies where possible, it is equally important when necessary to show children how their strategies may let them down (e.g. Committee of Enquiry into the Teaching of Mathematics in Schools 1982, Williams and Shuard 1976) and help them understand what the alternatives are.

THE USE OF CONCRETE MATERIALS

Using concrete materials in teaching number concepts and skills has come to be taken for granted as a method that will help children to learn mathematics in a more meaningful way. Their general acceptance has followed on the importance placed in the Piagetian stages in the development of children's thinking, of the physical manipulation of objects or the actual physical enacting in some way of mathematical operations such as addition or subtraction. Some of these materials are what are sometimes referred to in schools as 'junk' materials which include a wide range of everyday objects for use in counting and weighing activities in the classroom. Others are 'structured', i.e. they have certain mathematical properties built into them, as in the case of rods of different unit lengths or shapes divided into fractions. The tangible properties of the physical objects help pupils to store up mental structures which contribute to the development of mathematical concepts and their understanding.

The value of the use of structured materials in mathematical learning has more recently been called into question. Gravemeijer (1997) suggests that the tangibility alone of concrete apparatus of this kind does not equate with making mathematical sense of the objects. The mathematics inherent in the objects is not obvious to the child and 'passes over the situated informal knowledge' that they bring to the learning situation (p. 315). Using materials of this kind is not seen as a 'bottom up' process that begins with the learner (*ibid.*). Rather, the mathematical properties to be taught are inherent in the manipulatives and are built in for the purpose of eliminating any environmental features which may detract from the mathematics to be learned.

Research has shown that even where children are helped to master concepts through the use of manipulatives, they are not able to apply the concepts in problem-solving situations (Schoenfeld 1987), and that children do not develop an insight into the concepts being taught using these materials (Resnick and Omanson 1987). They may, for example, *observe* while using structured rods to learn about number that number has something to do with growth and 'big' and 'small', but they do not necessarily relate this growth to the intended number concepts. Similarly, Cobb (1987a) argues that manipulatives do not in themselves *convey* knowledge and that the mathematics within them will only be seen by those who already have the mathematical concepts. The research stresses that in everyday situations children are able to take hold of some aspect of a situation to develop a strategy to solve a problem with or without aids of this kind, and this kind of structured embodiment is not necessary to such learning. Carpenter and Moser (1984) have found, for example, that children are able to solve word problems involving mathematical concepts which, when presented out of any context and in a 'pure' mathematical format, they were unable to tackle successfully. In what is seen as a 'top down' approach, the concretized formal mathematics is 'given' in structured materials and no attention is directed to application or the way the mathematics is embedded in real-life situations; application follows the learning of the formal mathematics rather than the mathematics arising in some form of reality. (We shall see further evidence relating to the use of manipulatives later in this chapter.)

Gravemeijer (1997) suggests three models that illustrate the difference between the approaches discussed above. The first model where manipulatives are used is generally referred to as the *information processing* or *structuralist* approach and is seen as having two levels only: formal knowledge and a model of that knowledge in the form of the manipulative. The second or intermediate model represents the *situated* approach where there are three levels: starting at the bottom with a situation in which informal knowledge is embedded, a model of the situation, then the formal knowledge (the 'given' mathematics). The realistic model is seen as having four levels: the situation at the bottom level, the 'model *of*' the situation next, then the 'model *for*' where the situation itself is left behind and numbers alone are dealt with; and finally, the formal level at which the standard algorithm is reached. From the research, Gravemeijer (1997) summarizes the three main shortcomings of manipulatives as aids to children's mathematical learning:

1. the use of manipulatives does not in itself necessarily allow children to develop an insight into the mathematical concept being taught in the sense that they understand it;
2, even where children may have grasped the concept, this does not necessarily mean they know how and when to use it in a problem-solving situation;
3. the use of manipulatives does not reflect the children's own experience and ignores the informal knowledge and strategies they already have.

The mathematics in, or of, any objects used in teaching mathematics has to be mediated from the concrete to the abstract, and the child needs help in achieving this. Teachers have to devise ways of linking the situated informal knowledge of children, i.e. the experiences of number and their learning that is 'situated' in their general personal experience of the world outside the classroom, with the mathematical concepts it is intended for the children to learn.

THE FOUR OPERATIONS

The studies reported below illustrate different ways in which cognitive perspectives have influenced approaches used in researching children's work with number.

Addition and subtraction

Foster (1994) investigated children's difficulties with what appear to be simple addition tasks. He refers to the stages identified by Fuson (1992) in performing addition:

1. *count-all*, where each set is counted separately then the two are counted together;
2. *count-on*, where they count on from the first set through the second set to find the total;
3. *count-on from the larger*, where they reverse the order of the numbers and count on to include the smaller;
4. *count-on from either*, when they reverse the order of the numbers regardless of their size.

Children use these counting strategies together with *known facts*, which are *derived* to produce 'new' facts. The process is seen to be one of compression as the children reduce the number of steps in the process of adding.

Three different types of addition tasks were given to a sample of 147 children aged from 5 to 8 years old from Years 1, 2 and 3 in two United Kingdom primary schools. An example of each of the different types of presentation of the tasks is shown below.

1. Missing total $2 + 3 = \square$
2. Missing addend $2 + \square = 5$
3. Missing augend $\square + 3 = 5$

The results show a high degree of accuracy across the years for the first type of task (82%, 89% and 97% respectively). In Year 1, only 37% got the second and third types correct; in Year 2, 70% and 65% got the second and third types correct respectively, and in Year 3, the proportions were 90% and 83% respectively. This data was broken down further to group the achievement of individuals into three categories of high, low and medium attainers, and interviews were carried out with 24 selected pairs. Foster concludes that 'Once children were competent at all the available procedures they select one which best fits what they require' (*op. cit.,* p. 366). He states that:

> When actually solving a particular example children resort to the whole battery of skills and knowledge at their disposal. The left-to-right reading of a number statements is quickly over-ridden by the convenience of *counting on from larger* or *counting on from either* for the majority of children. As procedures these are probably more important than count-on itself. This study demonstrates that the required versatility is present in large measure in the successful group, and is not entirely absent from the less successful group. (*idem,* emphasis added)

The performance of the sample from one year to the next showed that the bottom third of each class showed increasingly better performances, although there was marked difference in the average of performance on all three tasks among the three groups. This further supports the notion of the proceptual divide (Gray and Tall 1994, Gray *et al.* 1997) which indicates 'a qualitative difference in performance between flexible use of number concepts on the one hand and counting procedures on the other' (Foster 1994, p. 366). Where proceptual thinking had not developed, children were able to fall back on counting procedures to reach a solution.

Gray (1994) explored children's addition and subtraction of two-digit numbers and found three levels of performance in 7- to 8-year-olds, where they moved from limited number facts and limited procedural competence (Group 1) to methods using excessive use of counting strategies (Group 2) to the top level of achievement where there was a spectrum of performance characterized by the flexibility in the methods used (Group 3). He found that two kinds of information processing were evident: one based on rearrangements of numbers and the other based on the use of instructions. Where the tasks (both in addition and subtraction) were presented horizontally (e.g. 24 + 43) children visualized rearrangements of individual numbers, but when tasks were presented vertically, they fell back on the use of algorithmic methods within this process. In the case of visual rearrangements, children made use of their knowledge of pattern and evidently did not feel constrained to complete the tasks using any one particular method. When presented vertically, however, they exhibited 'rule-like' behaviour which they clearly did not always understand. The reversal of success when moving from horizontal to vertical presentation is one of the more interesting outcomes of this study and is evidence of the results of constraints placed on flexible thinkers when they feel they must proceed according to given rules.

The kind of flexibility children can exhibit when given the freedom to do so is illustrated in an example used by de Lange (1996) from a study reported by Hiebert *et al.* (1996). The following problem was given:

> Find the difference in the height of two children, Jorge and Paulo, who were 62 inches tall and 37 inches tall, respectively. (de Lange 1996, p. 96)

One child counted up from 37 to 62, by ones (recorded by dots) and tens (recorded by 'sticks') so that it looked like this:

37 . . . > 40
40 / / > 60
60 . . > 62

Another child used the 'empty number line' (Gravemeijer 1994) and produced the following:

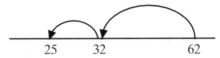

This child chose to subtract, started at 62, took away 30 to arrive at 32, and taking away 7 more left him at 25. The point Lange illustrates and emphasizes is that children should not only have a variety of models available to them to help them to carry out number operations, but they should be allowed to use their own strategies within those models.

Children's strategies in computational problems were studied by Ruthven (1998) who examined their use of mental, written and calculator strategies. The sample included one group of schools that had been involved in a 'calculator-aware' progamme for a number of years and had been encouraged 'to develop informal methods of calculation and to use calculators to explore number and execute demanding computations' (Ruthven 1998, p. 21). The second group of schools had not been involved in the programme and had a very different tradition related to how number was taught. Children from the calculator-aware schools were found to be more likely to use mental calculation as a result of having been encouraged to use and refine informal methods which involved (amongst other strategies) breaking numbers down (e.g. 45 to 40 + 5) as opposed to setting them out in columnar fashion with its stress on place value. Children from the remaining schools had a model of mental calculation which involved either doing a calculation quickly or abandoning it and exhibited a lack of fluency and flexibility with number. Ruthven draws attention to the strong implications this holds for some current approaches to mental calculation and numeracy, not least of which is the knowledge of number facts necessary to develop such fluency and flexibility. The message is that stressing an understanding of place value in early number work may not be the best way to achieve this.

Multiplication and division

In the early stages of mathematical operations, diagrams are often used as a way to help children to conceptualize the process. Although we may think that children can produce diagrams that reflect the concept being learned while they are in the process of learning it, it is not quite so straightforward. A study by Alseth (1998) supports earlier research (e.g. Fischbein 1987, Hembree 1992, Greer 1992) which shows how children will use diagrams in problem-solving situations, but are not able to do so unless they already have an understanding of the concept embedded in the problem. Alseth looked at children's perceptions of multiplicative structure within problems and how they used diagrams to represent this structure.

Sixteen 8-year-olds were given selected tasks involving multiplication and place value, each of which had a visual component. None of the children had previously been taught multiplication. The tasks were read out to the children and they were then presented with a selection of diagrams, each of which was relevant to the problem in some way, and asked which, if any, might help them to solve the problem. An example of the kind of problem given was if fifteen children were to sit three at a table, how many tables would be needed. Results showed that children selected diagrams that were not related to the multiplicative structure, but that they depended on counting to solve the problem. In the example given, it did not matter to the child how the figures representing children in the problem were grouped, as long as the total came to 15. The number of objects was the dominant feature in selecting a diagram and children relied on their counting skills. Because they had not yet been taught multiplication, they could not relate any of the groupings in the diagrams shown them that would correctly represent what happens when we multiply; they could not visualize the solution to the problem in terms of collecting equal sets or groups. As Fischbein says, 'A diagram is necessarily a post-conceptual structure' (1987, p. 158). To be effective, diagrams should be used alongside the teaching of a concept to help pupils to reach an understanding of the concept. As is the case with structured manipulatives, children do not necessarily recognize the mathematics in somone else's representations when they are in the process of learning them for themselves.

Ruwisch (1998) explored 7- to 9-year-old children's ability to carry out multiplicative problem-solving strategies using real-life situations. Three problems were given, one at a time and orally, to the children over a period of two to three weeks and involved:

1. numbers of goods in a pack;
2. the numbers of bottles of a given volume needed to produce a given number of glasses of juice; and
3. the number of tiles to cover the floors in a dolls' house.

She found that very few of the children used addition facts and either counted or used multiplication facts; and they were consistent in the strategy they used from one problem to another. A point noted was that when a strategy apparently had failed (for example where there was a remainder when trying to calculate the number of tiles), the children would invoke the real-world context being used and comment to the effect that they could probably go to a shop to buy different sizes that would fit.

A study to investigate when children reach an understanding of commutativity in multiplication was carried out by Schlieman *et al.* (1994). Previous studies were noted in which children who were street vendors in Brazil relied on repeated addition in performing calculations in the course of their selling (Schlieman and Carraher 1992, Nunes *et al.* 1993). A sample of 72 schoolchildren aged from 6 to 9 years old and 44 street vendors aged 9 to 13 years old were interviewed and given tasks to perform. The difference in age between the groups was accounted for by the fact that the older children came from poorer areas, and while they may well have attended school from one to four years, they would not have received regular instruction in multiplication. It was found that the schoolchildren had learned about, and could use, the commutative law. The majority of street vendors, on the other hand, appeared not to have moved from the stage of using repeated addition through to any understanding of commutativity. It is suggested that 'The everyday practice as street seller where numbers most often are used to refer to physical quantities has the advantage of preserving the meaning of the operations performed throughout the series of computational steps until a final answer is reached' and this 'really matters' in order to be certain the proper price is reached (Schlieman *et al.* 1994, p. 216). The children had continued to use the strategy of repeated addition because it proved reliable.

Gravemeijer (1997) draws on a study by Dolk and Uittenboogaard (1989) to illustrate realistic mathematics education in action in which the first steps towards long division are being taken. A class of 8- and 9-year-olds were told that 81 parents would be coming to the school and six parents would be seated at each table. They were asked to find how many tables were needed and the only lead given by the teacher was a drawing of a few tables. Some of the children solved the problem by repetitive addition and added $6 + 6 + 6 + \ldots$ until they reached 81 and all the parents were accounted for. Some began with multiplication: $10 \times 6 = 60$ and continued by further multiplication or addition. One child who knew $6 \times 6 = 36$ began with that knowledge and built up by doubling it to get $12 \times 6 = 72$, then added 6 and a further 6 to account for all the parents. Note the degree of realism in this problem where the children are expected to reason that they end up with 84 places for 81 parents, but they still need to seat the 3 who would be 'left over' with nowhere to sit if they did not put out that other table and have spaces to spare. The children discussed the solutions and were given a further problem of the same sort in the same context, i.e. finding how many pots of coffee would be needed for the parents if one pot gave 7 cups. Most children this time went straight to multiplying by 10 and only one used the repeated addition method. This illustrates how the approach involves the children in a problem and allows them to discover strategies at their own level, build on their knowledge and gradually to refine their strategies.

Steinberg (1994) explored the strategies and algorithms used by children in a Cognitively Guided Instruction (CGI) Grade 4 class, in carrying out division. The approach offered the children opportunities to solve a variety of problems, and to construct, present and discuss their strategies. Steinberg refers to work focusing on children's procedures and algorithms by Carpenter, Fennema and Franke (1993) which indicated that children 'usually have a better conceptual understanding of them and avoid errors that are common when children learn procedures in a rote fashion' (Steinberg 1994, p. 305). She gives the following résumé of the steps in learning to carry out division:

At first children model the semantic structure of the problems closely, with manipulatives. Thus they model partitive and measurement problems differently . . . Later children use more flexible counting strategies that are abstractions of their modeling strategies. To solve measurement problems, children can skip count or add or subtract until they reach the given number and then count the number of counts. In partitive division they use similar strategies based on guessing and adjusting the number needed in each group. Children then use derived facts strategies which usually exploit or involve some known multiplication fact, and counting on or adding. Most of the strategies are conducted orally, sometimes with the help of tallies to keep track, or writing the sequence of number in the pattern. (*ibid.*, p. 306)

A case study approach was adopted in which a teacher was interviewed thirteen times over the course of 30 lessons, pupils were interviewed and their solution strategies recorded. No textbooks were used and the standard algorithm for division was not taught. The different examples of children's strategies identified in the study were found to reflect those in the sequence given above. One such example involved the methodology of Pam who had set herself the task of dividing 9,786 by 14. Having already established that the use of multiples of a number for subtraction was more efficient than strigthforward repeated subtraction, she managed the task successfully. Steinberg concludes that by developing and discussing with them the personal strategies they used, the children were led to more meaningful learning of the processes in carrying out division. With respect to Pam, she says, 'Talking, explaining, answering questions and writing about her strategy in words and in symbols, helped her develop symbolic representations and meta-cognitive processes to explain her thinking' (*ibid.*, p. 310).

Murray *et al.* (1994) report findings of the third stage of a study of children's development of the concepts of multiplication and division using a problem-centred approach. Children were given two problems, one requiring multiplication of two two-digit numbers and one requiring division of a three-digit number by a two-digit number. The focus of the research was on the strategies used by the children and the possible effects on these of the problem-centred approach by which they had been taught since first entering school. They found that the children were still using their own strategies and had 'maintained their "power of thought" attitude and generally that the culture of the problem-centred classroom still prevailed, with extensive and flexible use of strategies and their knowledge of multiplication facts (*ibid.,* p. 404). The three schools in the study each displayed marked differences in their preferred strategies, a factor that was put down to the effect of the community environment engendered by the whole approach to learning, as opposed to the culture of the individual that might prevail in other classrooms. While the dominant strategies tended to be the same within a classroom, there was apparent variation within a given strategy that the researchers considered was determined:

(a) by the pupil's level of the concept of number; and
(b) by the number facts they knew and could call upon.

An example is given where the strategy was partitioning the dividend and one boy partitioned 784 into 640 and 144, while another used 320, 320, 128 and 16. Murray *et al.* state that 'This is one of the central issues in a successful problem-centred classroom: that students operate at the levels at which *they* feel comfortable' (*ibid.,* p. 405, authors' italics). They suggest that their pupils are aware of the transformations they make and *why* they make them, and that they therefore understand the nature of computational algorithms at a metacognitive level' (*idem*).

Anghileri (1995) has studied the importance of language in the children's learning of division, how they read and interpret the division symbol and the nature of the difficulties they might experience in doing division. The sample included Year 6 (10- and 11-year-old) children who performed certain division tasks, and were encouraged to discuss what they did and were taped as they did so. Six different ways of interpreting 12 ÷ 3 were identified:

(a) 12 divided by 3;
(b) 12 divided into 3;
(c) 12 divided into 3s;
(d) 12 divide by 3;
(e) 12 divide into 3;
(f) 12 divided up into 3s.

These were only the expressions involving the word 'divide' but there were many others using the words 'shared', 'split', 'how many', and 'grouped'.

Anghileri observed the teacher interacting with these children in a classroom situation where it was noted that time constraints seemed severe and not much appeared available for discussion, explanation or negotiation. She concludes that:

> Where there is evidence that children's learning is so clearly related to the language they use to interpret the symbols of arithmetic, teachers and researchers need to reflect on the ways that the classroom interactions may facilitate these two processes in order that true understanding will result. (p. 120)

Newstead (1996) reports on a later stage of this project and reinforces the need for greater negotiation within the classroom. Her focus was on the strategies the children used and noted that in some cases, children lost track of the strategies with which they began to solve a problem which led to an incorrect response, while others occasionally reversed the numbers due to misconceptions on their part (e.g. $6 \div 12$).

Fractions

Behr *et al.* (1992) state that 'There is a great deal of agreement that learning rational number concepts remains a serious obstacle in the mathematical development of children' (p. 296). Some of this difficulty may be due to the fact that the idea of a fraction is one of the earliest abstract ideas with which children have to cope since there is no 'natural' context in which they automatically arise (Booker 1998). There is not, for example, a situation involving fractions that parallels the kind of everyday situation in which children find themselves involved in counting. It is generally accepted that a lack of successful development and understanding of fractions in the earliest stages of children's learning will result in difficulty that will follow them through to secondary years of schooling. Such a lack can, for example, undermine their understanding of algebra since the rules that govern the manipulation of fractions are also important in solving many algebraic problems and equations.

Four constructs identified in connection with rational numbers noted by Kieren (1988) and reported by Behr *et al.*:

1. measure;
2. quotient;
3. ratio number;
4. multiplicative operator.

They suggest that the difficulties inherent in the teaching of rational numbers and fractions may arise from the fact that, when applied to real-life situations and looked at from a teaching point of view, fractions 'take on numerous personalities' presumably influenced by these four constructs (*ibid.*). They examine two of Kieren's contructs, the *part-whole* construct and the *quotient,* both of which accommodate an approach to teaching fractions in one of three ways:

1. as dividing a continuous quantity into equal quantities (e.g. capacity or length);
2. as dividing a number of objects into equal sets (e.g. twelve apples divided into four groups of three);
3. as dividing a single whole into a number of equal parts (e.g. a pizza into four equal portions).

The part-whole construct leads to two different interpretations of fractions and ¾ is used as an example to illustrate this point. The ¾ may be seen as *parts of a whole* where the three-fourths is three lots of ¼ *parts of a whole*; this can apply to both discrete and continuous notions of quantity and the 'unit' is the one quarter. Alternatively, three-fourths can be seen as a *composite part of a whole* where it represents one '¾ unit' and is linked with discrete quantities (as in the case of a single ¾-sized piece of one apple pie).

The quotient construct is interpreted in terms of both kinds of division (partitive and quotitive) and takes into account the different meanings the numerator and the denominator can have; it is also used in the application of fractions involving discrete or continuous quantities. Behr *et al.* (1992) provide a complicated model of the different constructs that are involved in the various possible interpretations of rational number, and the model in itself is evidence of the complexities involved in the notion of this area of number work. They not only provide a model of these two constructs but also a model of a fraction as an operator which provides a very detailed map of the possible pathways that can be taken when carrying out a calculation using a fraction.

A less formal and, as a result, more accessible consideration of the teaching and learning of fractions is offered by Streefland (1991a). He describes five approaches (see below) to the teaching of the concept of fractions and arrives at what he considers to be some general principles that should guide their teaching. The five, in a sense, map a route from an entirely formal approach of teaching through to one which is based on, or derived from, real-life situations.

The *structuralist* approach is the first and it is seen as a learning theory that views mathematics as 'cognitive acquisition, as an ordered, closed, deductive system' (*ibid.*, p. 15). Where fractions are concerned this means an emphasis upon equivalence and the rules and properties for the conversion of fractions with different denominators to equivalent fractions with the same denominator. Thus the equivalence relationship is powerful in this approach.

The *mechanistic* approach simply engages pupils in a sequence of phases in the learning process where the final goal is 'realization of the final algorithmic level' and application is limited. Within this approach Streefland comments that 'Many individual learning processes give the impression that the students are rowing upstream against a current of problems void of any context' (*ibid.*, p. 17).

Pragmatic structuralism is an approach that assumes that 'progress in the mathematical system will at some moment require the application of fractions' as they are brought into play in a vertical progression within the subject. Having gained a concept of fractions in earlier stages, the rules that govern their use as fractions are used in later mathematics as, for example, in algebraic operations.

The next approach is what Streefland calls *Intermezzo,* a term which encapsulates a combination of the structural approach with an attempt to make fractions more meaningful by the materialization of the concept in various formats. ('Materialization' here means the use of structured materials in the teaching of fractions.) This is undertaken 'in the service of vertical mathematization' but 'This materialization acquires thereby an artificial and arbitrary quality, because of the *a priori* presence of the intended goals' (*ibid.*, p. 18). It is argued that this presumes an insight on the part of the learner that, essentially, should be the goal or purpose of the learning. To take an example, a circular disc divided into four equal parts to represent quarters presumes that a child knows that, because it is one of four equal parts, it is a quarter. In itself, it does not help the child to arrive at the *meaning* of a quarter. At this stage, Streefland sums up the structuralist influence by suggesting that it ignores reality as a source of learning and 'material for producing a system is replaced by material for reproducing a system' (*ibid.*).

Finally, the *realistic approach* is seen as producing *'the insightful construction of the system'* (as opposed to the reproduction of it as in the foregoing approaches) (*ibid.,* p. 19). The five educational principles involved in the realistic approach were given earlier in this chapter and, in the adoption of a realistic approach, Streefland argues that:

> The concept of fractions and operations involving fractions are then regarded from many different points of view and various aspects are considered. Divergent contexts and processes are explored which can produce fractions, such as fair sharing, division, measurement, refinement, making mixtures, combining and applying recipes. But attention is paid, as well, to the various characteristics displayed by fractions in these contexts and processes, such as the fraction in the division operator, the fractions in the division relation, and so on. (*ibid.,* p. 21)

In other words, the realistic approach does not shy away from dealing with the various mathematical constructs identified with fractions noted by Behr *et al.* (1992) above. Aspects of this approach are reflected in some of the studies which we shall now consider, as well as some where a structuralist approach is more prevalent.

Teaching and learning fractions at primary level
Booker (1998) reports an ongoing study with a class of Year 4 children in Australia who had been followed through from Year 3 where they were

introduced to the naming conventions for fractions. This stage of the study adopts an exploratory approach with children working in small groups using investigation tasks and games 'designed to bring out issues in the modelling, naming and renaming of fractions' and by building on their intuitive knowledge of fractions (*ibid.*, p. 129). It is argued that the use of the unit fraction (e.g. $\frac{1}{2}, \frac{1}{4}, \frac{1}{5}$ etc.) as an introduction is confusing to children since they tend to generalize from the fact that 4 is greater than 3, hence $\frac{1}{4}$ must be larger than $\frac{1}{3}$. They confuse the 'fraction name' with the number of objects in a collection as, for example in the case of $\frac{1}{3}$ of 6 where almost half the children in the study said it was 3. This led Booker to adopt the convention of '1 one' in the study, because 'An emphasis on partitioning 1 one rather than taking parts of a whole lays a foundation for fractions to be seen as numbers in an extended number system' so that ultimately 'children can come to see that 1 one provides the basis of all fractions' (*ibid.*, p. 130). Part of the study involved playing games with fractions; when new fractions occurred (such as $\frac{7}{8}$), children had to name the fractions and 'there was a need to relate the class suggestions to the mathematical conventions that had arisen over time' as, for example, with 13 eighths being written both as $\frac{13}{8}$ or 1 and $\frac{5}{8}$ (p. 133). It was found that by moving from a part/whole approach to the '1 one' concept of fractions, confusion in naming improper fractions and mixed numbers was reduced and allowed the 'insightful reconstruction' with respect to fractions to take place (Streefland 1991).

Alvarez (1994) reports on a case study of a Grade 4 class of children in a Mexican elementary school. The focus was on language and representation in the addition and subtraction of fractions. Children were given tasks such as 'Invent a problem that contains: $\frac{5}{8} + \frac{3}{4}$.' Some of the tasks included a pictorial representation in which the children had to indicate the remaining fraction of a whole by shading, but in cases where there was no pictorial representation it was found that children provided their own which occasionally led them astray. The difficulties that can arise in using illustrations with fraction work have also been explored by Nickson (1997). She found that while most children in a study were able successfully to identify one quarter of a pizza in a drawing, considerably fewer were able to identify one quarter when the illustration was of a race track represented by a long thin rectangle. There was evidence that some children could connect the pizza with the face of a clock and establish a quarter in that way, whereas there was nothing in their immediate experience with which could easily connect a thin rectangle.

The use of physical objects to represent fractions provided the focus for an investigation into the development of Grade 4 children's thinking about fractions undertaken by Arnon *et al.* (1994). Their study was based on the Theory of Learning Systems of Nesher (1989) which is built on the Piagetian notion that children learn new concepts by reflecting on the physical handling of objects. Nesher's theory is composed of three domains:

1. the *knowledge domain* consisting of the mathematical concepts involved;
2. the *exemplification* domain made up of the physical objects to be used;

3. the *activities within the knowledge domain* which is the stage at which the child no longer touches the physical object or perhaps even needs to see them.

The study set out to investigate how the shift from the second to the third stage happens. The researchers also focused on the idea that the learning of new concepts begins with the interiorization of an action which has two characteristics:

1. the learner has the ability to perform the required action in her or his head;
2. the learner is aware of her or his action and in some considerable detail.
    (Dubinsky 1991)

In their study, two classes of children formed the sample; one was taught using the Learning System as described above (using booklets, prefabricated physical manipulatives and the algorithm for multiplying unit fractions by an integer), and the other was taught by a combination of the Learning System and Dubinsky's idea of interiorization (with no algorithmic teaching). In the latter case, the pupils had, at some point in their learning, actually constructed their own physical representations of fractions including representations of non-unit fractions (e.g. ¾). The aim of the teaching unit was for the pupils to learn how to multiply a fraction by an integer. While it was found that both classes achieved a degree of interiorization, those in the second class achieved the most. In their case, where greater emphasis was placed on learning via actions that could be performed in the learner's imagination, 'more children interiorized the multiplication of a fraction by an integer to a degree where they became independent of the concrete objects' (*ibid.,* p. 39).

A different perspective was used by Maher *et al.* (1994) in investigating children's thinking about fractions. A class of 25 9- to 10-year-olds were observed and videoed in their classroom situation as they did tasks involving fractions. The constructivist approach adopted in the study is described as follows: 'In studying the development of mathematical ideas, problematic situations are posed and students are observed building solutions, developing arguments, constructing models, comparing models, discussing their ideas and negotiating their conflicts' (*ibid.,* p. 209). The teacher's role is to facilitate, and the teaching style is seen to be one of guiding children's thinking. The task described involved the children in placing the fractions ½, ⅓, ¼ . . . ⅒ in the interval between 0 to 1 on a number line. They were also asked to place ¾ where they thought it should be. Some of the children placed the ⅓ on both the ⅓ position and the ⅔ position. One pupil argued that the ⅓ could be placed anywhere on the line and modelled the situation using Cuisenaire rods using three of the red '2' rods against a green '6' rod to illustrate his point. This was in contrast to another pupil who saw the thirds in terms of a cumulative structure which started at zero and progressed from there. The first pupil gradually began to make the distinction between the 'numbers which define the *intervals* and numbers which define *positions* of the intervals' (*ibid.,* p. 213, italics added). The way in which this development took place within the classroom is seen by the researchers to be an example of the importance of allowing children to reconstruct their thinking so that their learning is meaningful (Streefland 1991a).

Three main factors obstructing an understanding of fractions are identified by Newstead and Murray (1998) from the research (Baroody and Hume 1991, Streefland 1991, D'Ambrosio and Newborn 1994) as follows:

1. the way and the order in which content related to fractions is presented, in particular using only halves and quarters and using concrete apparatus that has been divided into fractions;
2. classroom environments that do not allow informal everyday conceptions of fractions to be formed and misconceptions go uncorrected;
3. inappropriate ideas related to whole numbers which lead to fractions being identified in terms of the whole numbers used in their representation – i.e. numerator and denominator are seen as separate numbers.

Newstead and Murray (1998) set out to explore children's understanding of fractions and operations with them with 370 children from Grades 4 and 6. The children were given written tests to evaluate their understanding in which some of the items were contextualized and some context-free. It was found that children used drawings of shapes divided into fractions reflecting the way they had been taught them in school, and this caused difficulties with those children who had not been taught fractions of continuous amounts. Children were also found to rely on a fixed approach to calculating equivalent fractions without being able to reason it through, and this sometimes let them down when they multiplied either the numerator or the denominator (but not both) by the same number. Difficulties also arose from misconceptions children held from incorrect intuitions about fractions, and this was evident when they were asked $2 \div \frac{1}{2}$ and they could not read it as 'How many halves are there in 2?'. This was seen as the probable result of insufficient exposure to a variety of situations involving fractions. A major difficulty the children had was in viewing the component parts of a fraction as two whole numbers and not seeing a fraction as an entity in itself (as shown when adding two fractions, adding the denominators to each other and numerators to each other). This was also demonstrated when comparing the size of fractions and children were misled by thinking larger denominators gave larger fractions.

The results of the study indicate that in common with other research, these children have 'after their first few years of school, limited and *limiting* understandings of fractions which persist into the upper elementary grades' (Newstead and Murray 1998, p. 301). On the positive side, it was found children could 'use non-mathematical interpretations to make sense of unfamiliar fraction situations in the context of problems' (*idem*). This is seen to reinforce the findings of previous research where conclusions are that the introduction of written symbols and algorithms with fractions should not take place until children have experienced fractions as single quantities (Baroody and Hume 1991, Empson 1995).

A secondary-level view of fractions
A study which considered older pupils' conceptions of fractions was carried out by Pirie *et al.* (1994). The focus was the images 15-year-old pupils had of fractions, and the researchers state that 'Image making means the performing of actions by the learner, actual or mental, to get some idea of the concept under consideration' (p. 247). The point is made that two kinds of image of fractions are necessary to understand operations with fractions that are important to an understanding of algebra. The example is given of twelve people sharing half a pizza as making sense, whereas the idea of 'half a person sharing twelve pizzas does not' (*ibid.,* p. 248). Part of the rationale for this study involved the idea of 'Cycles of Responsibility' where the evolution of the responsibility in helping pupils to achieve a meaningful understanding of fractions involves the teacher in 'folding back' and leading pupils to draw upon their previous learning that is relevant to the immediate situation. In order to probe pupils' grasp of the concept of fraction, they were given a questionnaire with six questions relating to fractions which were deliberately left open-ended. Four major images emerged:

1. *Division* – e.g. 'A quantity – numbers or sometimes a letter – divided by another quantity.'
2. *Part of a whole* – e.g. 'A fraction is a section of a whole thing.'
3. *Number* – 'It is a number that has not been put into a decimal.'
4. *Way of writing* – e.g. 'A number over another number.' (*ibid.*, p. 250)

Classroom discussions are reported as evidence of the way in which the teacher can 'fold back' and help pupils to conjure up earlier images and by recapitulating in this way, to come to an understanding of the new learning with which they are involved. The authors point out that the fourth image given above is a powerful one when considered in algebraic terms as a/b. The outcome of the study indicated that the 'part of a whole' was a strong image for many and the teacher used this to advantage in her explanation of algebraic expressions such as $2x^2/3y + 4$ and evoked 'a broader image for rational numbers which was crucial to more sophisticated, algebraic working' (*ibid.,* p. 253). A general conclusion drawn from the study was that the pupils who had a multiplicity of images, and could move flexibly back and forth amongst these images, had least difficulty in understanding the new algebraic work.

Proportional reasoning in solving rate problems
In a study which set out to explore the character and effectiveness of children's proportional reasoning in solving rate problems, contextual problems were presented to children from different cultural backgrounds (American and Israeli). Children were participants in the Connected

Mathematics Project (CMP) and came from Grades 6, 7, and 8. Their mathematics curriculum was structured in such a way as 'to develop students' knowledge and understanding of mathematics that is rich in connections – connections among the core ideas in mathematics and its applications' (Ben-Chaim *et al.* 1998, p. 247). The curriculum was organized around 'interesting problems settings – real situations, whimsical situations, or interesting mathematical situations' (*ibid.*).

The idea was to get children to observe pattern and relationships and to engage in conjecturing, hypothesizing, verbalizing and generalizing these. Data was gathered from 124 children in eight Grade 7 classes with a control group of 91 children from six Grade 7 classes. The children were presented with five problems involving proportion, two with unit price, two with proportional relations of the distance/time/speed type and the fifth dealing with population density. They did the first four rate problems as part of a written test and were contextualized, while the fifth was purely numerical and was done by 25% of the sample in an interview situation. During the interviews, children were also questioned about their thinking with respect to solving the written test problems.

The results from the study showed that children involved in the CMP 'out-performed' those who were not by 53% to 28% on all the tasks (*ibid.*, p. 272). They were able to give meaningful explanations, both written and oral, of their work and demonstrated an ability to draw on a variety of strategies in arriving at these solutions.

There are two implications for teaching:

1. the power of the 'unit rate' approach to problems should be highlighted as one which children can use to generate solutions over a variety of problems;
2. the effects on children's ability of number structure and context familiarity to solve problems of this kind need to be taken into account.

In order to develop the concept of proportion and so to solve relevant problems correctly, children need to be given opportunities to approach contexts that are meaningful to them. The researchers suggest that both of these factors should be manipulated to good effect in helping children to develop their skills in proportional reasoning and it is suggested that work aimed at such development is essential in the middle years of schooling. Grade 7 children 'are capable of developing their own repertoire of sense-making tools to help them to produce creative solutions and explanations' and teaching should take advantage of this repertoire and 'extend it to the domain of proportional reasoning' (*idem*).

Understanding percentage
Percentage problems in which percentage is viewed in terms of multiplicative comparison formed the basis of a study with Australian children in Years 8, 9 and 10 (Cooper *et al.* 1998). Calculating percentage was seen in terms of the

three components of a straightforward multiplication task: the Intitial Quantity, the Comparison and the Goal Quantity. For example in $30 \times 5 = 150$, the Intitial Quantity is '30' (the number being multiplied), the Comparison is '$\times 5$' (the number by which the Initial Quantity is being multiplied) and the Goal Quantity is 150 (the result). This gives rise to three types of problem and different operations to solve them. When the Goal Quantity is unknown, the operation is multiplication; when either the Comparison or the Goal Quantity are unknown, then it is necessary to divide. These are known in the literature as the following:

Type A Percentage unknown – finding part or percent of a number, e.g. what is 25% of $60 (i.e. goal quantity unknown);

Type B Percent unknown – finding a part or percent one number is of another, e.g. what % of $50 is 15 (comparison unknown);

Type C 100% unknown – finding a number when a certain part or percent of that number is known, e.g. 25% of the cost is $15, what is the total cost (initial quantity unknown). (*ibid.*, p. 200)

The purpose of the study was to determine whether or not children actually use the multiplicative schema when they solve problems involving percentages for each of these problem types. The research suggests that the four main strategies used to solve such problems are:

1. *cases* where children classify the problem as one of the three types and then select their strategy;
2. *proportion,* where the problem is seen in terms of an unknown and the percentage is converted into a fraction with a denominator of 100;
3. *unitary,* where 1% of the Initial Quantity is calculated and then multiplied by the total percentage required; and
4. *formula,* where the solution is viewed as a formula (e.g. $P = QR$) and substitutions are made.

This study involved 90 junior secondary children who were classified as *proficient* (able to solve all types of problem), *semi-proficient* (able to solve Type A problems but not B or C) or *non-proficient* (unable to solve any other three types of task). Six children of each type were then interviewed and given tasks:

(a) to explore their schema of percent;
(b) to identify their strategies in solving percent problems; and
(c) to explore their use of visualization in the process of reaching solutions or by using diagrams if they had not already done so.

There was evidence of all of the three solution strategies being used by the proficient children (of whom there were only six of the total population of 90). They used a variety of strategies (beginning by classifying the problem), the semi-proficient used fewer strategies (proportion and formula) and the non-proficient who used the formula strategy only.

Proficient children were found generally to be flexible in their approach and to use trial-and-error methods in the process of reaching a solution, as well as appearing 'to be able to analyse problems in terms of their meanings, predict the required operation and gauge the size of the answer relative to the numbers they had been given' (Cooper *et al.* 1998, p. 205). Semi-proficient used these strategies only as a method for checking their answers. No children from any categories spontaneously used diagrams in solving the different tasks, but when asked to do so, all the proficient children were able to produce such drawings, while only one from the semi-proficient category and five from the non-proficient did so.

The main feature separating the proficient and semi-proficient children in this study was the fact that the latter could not identify the three types of problems involving percent. They lacked the flexibility of those in the proficient category, could not model a percentage problem using a diagram, and even though able to identify when an answer did not make sense, they were able to find an alternative strategy. The non-proficient children saw the superficial features within a problem as typifying it, for example whether it involved selling or buying, and attacked problems in a routine way, changing the percent into a decimal and using the formula or at times relying on key words. They had poor computational skills and could not identify whether an answer was sensible or not.

There are two implications from the study for helping semi-proficient and non-proficient children to achieve better in solving percent problems: firstly, that good computational skills are necessary, and features of these must be explicitly taught in the context of percent problems (e.g. number sense, approximation, estimation, conversion to decimals, and fractions). Trial and error strategies should be taught alongside these skills. Types B and C problems should be part of the teaching of percent problems and also of problems involving the multiplication of fractions and whole numbers, and the connections between all three should be explicitly made. The question of whether the multiplication schema that formed the focus of this study should be taught remains open to question.

Measurement

The topic of measurement within the mathematics curriculum was described in the report of the Cockcroft Committee (in the UK) in the following way:

Measurement is fundamental to the teaching and learning of mathematics because it provides a natural 'way in' both to the development of number concepts and also to the application of mathematics over a very wide field. (CITMS 1982, para. 269, p. 78)

While this places measurement in a realistic setting which provides a logical way in to teaching mathematics in real-life contexts in which number concepts are applied and used, measurement is in itself an exceedingly complicated concept. Carpenter (1986) notes that 'Young children can learn to perform accurately a number of simple measuring procedures and use the results to make judgements about the magnitudes of quantities' (p. 122). He goes on to say, however, that while such procedures are not meaningless and carried out in a routine way, they are limited because:

> Although young children can measure using single units, initially they are unable to establish relationships between units. The ability to establish such relationships is a milestone in the development of more advanced measurement skills. (*ibid.*)

It is tempting to treat measurement as a mathematical concept that may be more easily accessible to children than are many other concepts, simply because it exists all around them in their everyday lives. However, a consideration of some recent studies provides evidence of some of the persisting difficulties children have in this aspect of mathematics.

Establishing what measurement means
Children's progression in the learning of measurement is affected by a variety of factors which include the following identified by Brown *et al.* (1995, p. 160):

- the changing motivation and intellectual maturity of the child;
- the sequencing and nature of teaching and activity experienced, both within and outside the school;
- the logical structure of concepts within each discipline.

Their research set out to discover:

(a) whether there is an identifiable progression to be found in children's learning of measurement concepts;
(b) the extent of variation amongst children and across different age groups in this learning;
(c) what the relationship is between children's learning and the teaching they have experienced.

The study focused on length and weight because they were seen to have different qualities, the former being spatial (in the sense that it is possible to *see* length) and the latter being non-spatial. The project took place over two years and the sample included four age groups from Years 2, 4, 6 and 8 from two secondary and two primary schools in a London borough. This meant it was possible to carry out both a limited longitudinal study as well as a cross-sectional one. Twelve children from each year group were given two practical tasks (one relating to length and the other to weight) in interview situations, at the beginning and the end of the project. Observations of classroom teaching were also undertaken. The tasks involved four conceptual strands that had been identified in previous research (Dickson *et al.* 1984) and included units, number, scale and continuity.

The outcomes indicated that progression could be described from two perspectives. First, there appeared to be 'some aspects of measurement where development seemed to relate strongly to pupils' experiences whether in the home or in school (including science or home economics), but which seemed to be relatively straightforward conceptually and not dependent on other ideas' (Brown *et al.* 1995, p. 164). Pupils' awareness of units of length is given as an example where a few of the younger pupils, and almost all the older ones, knew the units. This was the area of understanding in which there was most improvement between the beginning and end of the study, although there was much less change in their understanding of the *size* of units. The second perspective of progression was 'across age, with a conceptual basis' and three such examples are given. One concerned number concepts where some children were not aware that aggregating the values of separate weights used was necessary to find the total weight of an object. This also related to measures of length where children did not realize they had to add the length of the ruler the correct number of times to arrive at a total when measuring a distance of about 1.1 m. Another concept in this category was the relationships between units where only six of the twelve children could identify a relationship between centimetres and millimetres. Finally, the third strand of progression was with respect to continuity and 'Whereas young pupils believed there was a unique answer but which they might not "get right" because of their own errors, older pupils appreciated that there were other sources of inexactness' (*ibid.*, p. 166). The researchers note that there had been little mention of this aspect of measurement during the teaching observed and suggested that the progression may have resulted from the interviews themselves. The overall conclusion drawn is that progression in learning in measurement is not clear-cut and that many strands are involved, not all of which necessarily are inter-correlated and some of which are very dependent on number concepts. They conclude that: 'Some competences do not seem to develop much with age, either because they are not much taught or because they relate to relatively fixed personal characteristics. Others are shown to be easily taught and seem to be more strongly related to experience' (*ibid.*, p. 167).

Pike and Forrester (1996) also carried out an age-related study of children's ability to estimate and their number sense, using measurement as the context. Laptops were used in this study involving 6- to 11-year-olds and the aims were:

(a) to assess children's number sense using three tasks;
(b) to assess their estimating ability using two tasks, one on length and one involving area.

There appeared to be no progression across the ages in the children's ability to estimate (although they were better at estimating length than area) but there was progression in their number sense. They conclude that 'whereas children's ability to use and perceive number relations and, secondarily, their understanding of relative number magnitude, appeared to influence their ability to estimate area, no such influence was found on their ability to estimate length' (*ibid.*, p. 47).

These results appear to support those of Brown *et al.* (1995) and reinforce the fact that the conceptual development of measurement is not easily mapped.

Linear measurement
A study carried out by Boulton-Lewis *et al.* (1994) into children's strategies in measuring length investigated their use of different kinds of units, standard and non-standard. Two theoretical perspectives were used to inform the research. Reporting the work of Halford (1993), they refer to his structure mapping theory and the levels of structure mapping identified according to the complexity of tasks. These are related to ages: *element mapping* level (1–2 years) where children function in a nominal way and they may use terms like 'big' when, for example, referring to length; *relational mapping* level (2–5 years) where they compare two lengths and can use a standard measure without necessarily understanding why it is structured the way it is or the units involved; and the *system mapping* level where they have cognition of three elements as well as the relationships between them. At the latter level they should:

(a) have the concept of invariance and understand that the length of something is unchanged whatever its position;
(b) understand transitivity when comparing three lengths;
(c) understand the procedures for measuring length that involve taking into account comparative length and the number of sub-units.

The researchers also took into account the recursive theory of cognitive development put forward by Case (1992) which is described in four stages and in which each stage is subsumed within the next to produce higher-level operations. These broadly follow the levels identified by Halford and the final stage is reached when children are able 'to construct and compare entities that result from numerical operations' and as a result should be able to use this understanding when measuring length situations which require comparison of lengths of objects in terms of numbers and units (Boulton-Lewis *et al.*, p. 129).

In earlier work carried out Boulton-Lewis (1987) predicted sequences of length measuring knowledge and identified three distinct types of task according to the variables within them:

(a) equality and inequality of length, ordering lengths by pairs and number knowledge related to these tasks;

(b) recognition of the invariance of length, transitivity and the conservation of length;
(c) measurement with a ruler, a strategy for measuring length using arbitrary units and the ability to reason transitively.

In this research Boulton-Lewis *et al.* found that 'children could succeed with using a ruler before they could devise a measuring strategy using arbitrary units and reason transitively' (Boulton-Lewis *et al.* 1994, p. 130). In the present study, it is argued that:

> In order to realize that arbitrary measures are not reliable a child must reconcile the varying lengths and numbers of arbitrary units and reason transitively. On the other hand to use a standard device such as a 30 cm or metre ruler to make pair by pair direct comparisons of lengths of objects is a less demanding task. It also has the advantage that it appears to be a real-world meaningful activity. (*ibid.*)

They note the conventional belief that arbitrary units should be used before standard units in order that children come to understand the need for a standard unit with which to measure. This widely used strategy is the one advocated in the Cockcroft Report when they state:

> Practice in ordering lengths, capacities and weight enables a young child to develop understanding of concepts such as 'more than', 'less than', 'longer than', 'shorter than'. After this the child learns *first of all to use non-standard units* such as handspans and cupfuls and then the standard units to measure continuous quantities. (CITMS 1982, para. 269, p. 78, italics added)

However, Boulton-Lewis *et al.* argue that the information-processing demand of the use of arbitrary units is considerable as well as unrealistic. With this in mind, they set out to explore what measuring instruments children would choose to use in a variety of situations.

The study was carried out with 70 children from Years 1, 2 and 3 and their teachers had to identify what objects were to be measured, the strategies the children would be taught and the materials they would use. Materials included a variety of arbitrary units such as unifix cubes, unmarked sticks, half- and full-length matches and string. Every child was interviewed in terms 2 and 4 with a sixteen-week interval between. The procedure then was to present each child with three measuring tasks, one involving pieces of string, another a skipping rope, and the third, matches in different end-to-end configurations. Children were invited to measure these with whatever they chose from the materials available. They found that where arbitrary units were chosen, there was 'a significant *increase* in *unsuccessful* use' in Year 1, a *decrease* in Year 2, an *increase* in early Year 3 and then a decline after that (*ibid.*, p. 133, emphasis added). They found that although 'the use of arbitrary devices increased in Year 2, probably due to teaching effects, that there was a high interest in the use of a standard device even in Year 1, that this increased over the three years and significantly in Year 3' (*ibid*). They sum up their findings by saying that 'there was a significant increase in understanding of the length measuring process over the three years' that matched the outline of development in the cognitive theories and, given the opportunity, 'children will choose to use a standard measuring device, if possible, even if they do not understand it fully or use it accurately' (*ibid.*).

Measurement of area

Nunes *et al.* (1994a) investigated the hypothesis that 'children's success in understanding area is not independent of the resources that are given to represent area during problem-solving' (p. 257). They argue that the measurement tools available for children have to be mastered and, in themselves, constitute part of the way in which children come to structure their actions and strategies. The study was carried out in three phases with 48 pairs of children 8 to 11 years old in multi-ethnic schools in London. The first phase probed whether children are more likely to achieve a multiplicative idea of area using bricks or a ruler and whether they could transfer what they learned in connection with rectangles to solving problems with other geometrical shapes. They worked in pairs and were randomly given either bricks or rulers. They found that more children using bricks calculated the area of shapes through multiplication than did those using rulers, and that this strategy was used again when they measured parallelograms. Success rates were low when both bricks and rulers were used to find the area of a triangle. Children using bricks referred to numbers of rows and columns in giving their explanations. Rather than using the calibrations on the ruler, children tended to find out how many times the ruler would fit over the shapes.

The second phase of the study (in which different children of the same characteristics were involved) investigated the results of teaching that relied on different tools, but the focus was different. The researchers wanted to find out whether children could transfer what they had learned from a demonstration of the comparison of the areas of two rectangles to the comparison of the area of another pair, whether they adopted the same procedure in doing so and whether their success was linked to the tools used in the teaching situation. A ruler and bricks were given to different pairs of children once again. In this case they found that 'although all children could repeat a procedure they were taught immediately after instruction, children instructed through an isomorphism of measures procedure were more likely to devise adequate solutions to new area problems' (*ibid.*, p. 261). Finally, in phase 3, the results of the outcomes of the first two phases were compared. The first group had essentially been involved in 'discovering' area whereas the second had been taught by means of a demonstration. The statistical comparison carried out showed that the only significant effect was that of the method used, whether or not the learning situation had been one of 'discovery' (phase 1 children) or whether procedures for how to find the area had been demonstrated to them (phase 2 children).

The researchers conclude that 'it is clear that one cannot analyze the cognitive demands of an area problem without considering the resource available in terms of problem representation to the problem solver' (*ibid.,* p. 262). The materials that are presented to children in the first instance in teaching them the concept of area affect their success in coming to understand it. It is also noted that although children are given the opportunity to tessellate figures when they are taught area, they usually are asked to *count* them. It is suggested that this leaves them without a problem to solve and they have no incentive to find some way of calculating area. The implication is that if these two matters were attended to in teaching strategies, the teaching of area might be more effective. No gender differences were noted in the analysis of data.

Owens and Outhred (1997) studied 7- to 10-year-old children's early concepts of area using tasks involving tiling different shapes. In this instance they were concerned to find out whether children understood what it was they were measuring when they measured area. They note that evidence suggests that in early stages, children have difficulty in choosing an appropriate shape for tiling and use tactics such as flipping and turning them (Mansfield and Scott 1990). They also quote evidence that while concrete experience helps to engage pupils' imagery, it should also involve pupil–pupil and pupil–teacher dialogue about the ideas arising from this kind of activity if the development of the mathematical concepts is to take place (Hart and Sinkinson 1988).

Two hundred pupils aged 7 to 10 years in five multicultural schools were given a spatial test using different shaped tiles with which to cover other given shapes. They were tested three times over a period of three months. It was found that the use of triangular tiles was more difficult than rectangular (including square) tiles and this was especially the case when the shape being covered was not itself triangular. The activities appeared to help pupils to visualize a grid of squares in some instances (e.g. when covering pentominoes). Any drawings they had done were analysed and this indicated that their strategies included tiling around the sides, starting from a corner and filling in towards the centre, drawing individual tiles in rows, and representing rows by marking them off on the sides of shapes. It was concluded from the study that the use of drawing in the development of area concepts helps children to develop abstractions and to recognize the units that go to make up a shape, and this supports the findings of other research (Wheatley and Reynolds 1996). While advocating the use of tiling in developing the concept of units in measurement, they note that there is a danger that structured materials such as tiles might reduce 'non-investigative area tasks to counting tasks' and they advocate greater emphasis on being able to 'transform shapes to other orientations, recognize and partition shapes, and identify key features of shapes' (Owens and Outhred 1997, p. 319).

The outcomes of this study have two things in common with that of Nunes *et al.* (1994b) reported above. In both,

(a) the children had difficulty in coping with the area of a triangle;
(b) the researchers recognize the problems inherent in repeated 'covering up' with unit objects as part of the methodology of teaching area.

Children tend to fall back on counting strategies and thus can be distracted from arriving at the rule for finding area in a meaningful way.

Concepts of area at secondary level
Kidman and Cooper (1997) investigated:

(a) the way in which pupils integrate visual stimuli when calculating area; and
(b) whether these rules of integration change with time.

In reviewing the literature, they found that, at about the age of 8, children 'are in a transitional stage between the additive and multiplicative rules' to calculate area and they set out to explore whether there was evidence of such development across three year groups. A sample of 36 children from Grades 4, 6 and 8 were given two tasks in which they had to make judgements about the amount (in terms of area) of bars of 'chocolate' as they appeared in three different situations. In the first, they had to judge the relative areas of sixteen pieces of chocolate with different dimensions corresponding to factorial combination of 3, 6, 9 and 12 cm. In the second task, a straight line 'bite' across a corner had been removed, and in the third, the 'bite' was semicircular from the side of the bar so that although the area had decreased, the perimeter increased. The children were also interviewed about how they had made their judgements, and their understanding of area and perimeter was probed. As with studies noted earlier (Brown *et al.* 1995, Pike and Forrester 1996) they found that the differences across the grades were not as clear as expected. They also found that 'The perception of area of [a] rectangle being related to the sum of the rectangle's dimensions is fairly constant across the grades' and they note that the Grade 8 pupils had not progressed much beyond the lower two grades (*op. cit.*, p. 141). 'Over 50% of the students (22 out of 36) changed their integration rules across experiments' and this was sometimes from using multiplication back to using addition (*ibid.*, p. 142).

This confusion often exhibited by children between the measurement of perimeter and area has been the focus of much of the measurement research. Comiti and Moreira Baltar (1997) also explored it using a case study method with 12- to 13-year-olds using paper-and-pencil methods as well as 'Cabri-géomètre'. In the first three stages of the study, the pupils were taught the stages of arriving at the formulae for finding the perimeter and the area of two-dimensional figures. In the final stage, they were tested using Cabri-géomètre software. At this final stage, although they knew it, 'they did not spontaneously resort' to the formula for area. They appeared still to be attending to linear measurements and thought that a parallelogram and a rectangle with equal sides had the same area (*ibid.*, p. 269).

IMPLICATIONS FOR TEACHING

There is considerable diversity in the studies explored in this chapter although, apart from initial theoretical perspectives, most have related to mathematics in the primary years. This seems inevitable since research related to number concepts and learning of operations tends to focus on the earlier years of schooling where these topics are the explicit content of the curriculum. However, many of these studies are relevant to teachers of mathematics at all levels because they help to identify the potential source of children's learning difficulties in later years. Misconceptions of basic concepts that persevere from early learning experiences are very often the cause of lack of understanding of more sophisticated mathematical concepts in later years.

Shifts in theoretical perspectives

1. *Messages about what mathematics is*
Shifts in both psychological and philosophical issues in mathematics education place an emphasis on the 'child in context'. It is important to recognize that, as teachers, we convey messages about the nature of mathematics by the way we teach it. The theoretical shifts suggest that if the message we want children to have is that mathematics is a 'doing' subject and an activity which is problem-oriented, then it needs to be taught in that way. The different perspectives offered in the research emphasize that the teaching of concepts should, in the first instance, be grounded in the reality of the children. They need to be put in situations where they have to mathematize. There are facts to be learned and procedures to be carried out which the research suggests that pupils will come to know and understand if there is a meaningful reason for doing so.

2. *The role of sharing and discussion*
Part of the process of coming to understand mathematics lies in sharing through discussion in the course of learning activities. This can lead pupils to the recognition that in mathematics, as in other subjects, different perspectives can be held and different strategies used. With development of these ideas, children come to realize the importance of consensus amongst themselves, for example about certain strategies and conventions, which will become more refined as they progress through the discipline. In the process of this kind of discussion it is possible to become more aware of pupils' thinking and their misconceptions, and to be able to guide their learning accordingly. Discussion allows pupils' thinking to be made explicit which in turn can help to build up a repertoire of their strategies that will be identifiable and can be useful in further teaching.

3. *Reflection and ownership of learning*
With the process of sharing comes reflection. If children are asked to justify to their peers why they have done what they have done, they are necessarily

having to think about it. This places some of the responsibility for their learning on them and they gain some ownership of mathematics in this way.

4. *The child's context and learning*
While emphasis in teaching remains child-centred, the perspective of what constitutes child-centredness has changed. Because children are viewed as 'constructors of knowledge', it becomes more important for teachers to know what they bring with them in terms of past experiences and informal knowledge, as well as their cultural context. An increased awareness of this kind could help to prevent unnecessary repetition of what children already know and, by acknowledging this, give credibility (where it is due) to this informal knowledge as opposed to reinforcing the idea that their personal experience is not meaningful to others. There is a delicate balance to be achieved here, however. While it is desirable to build on this learning, at the same time misconceptions cannot be allowed to persist and the kind of intervention that is necessary needs to be recognized.

5. *Numeracy and autonomy*
Current views of numeracy stress the *application* of number and not merely the knowledge of facts. It follows that teaching that emphasizes using number in meaningful contexts is important in order for children to become numerate. By applying number to solve different problems, they learn to transfer what they know from one situation to another and will, in the process, learn something about the structure both of kinds of problems and of the mathematics they use to solve them.

Number and counting

1. *Informal learning*
Much of the informal mathematical knowledge children bring with them to school relates to counting. Children's individual counting strategies need to be recognized and understood and built upon, either by being extended or gradually being replaced with more effective ones. This kind of knowledge can help to identify whether, for example, a child may be stuck at the 'proceptual divide' where they know particular procedures but cannot recognize when to use them appropriately. They have to learn to match the appropriate procedure to what is required in the problem being solved. The process of selection of this kind is fundamental to effective mathematical learning.

2. *The effect of teacher expectations*
Together with an increased awareness of the mathematics children already have, it is important to be aware of the power of children's perceived expectations of the teacher and the messages they may unintentionally convey. The research indicates that although children may come to school with some knowledge about informal number work, rather than being a power

base to build on, this knowledge can sometimes become devalued and children lose what confidence they may have had.

3. *Counting and cardinality*

Difficulty with number work often arises as a result of a lack of understanding of the cardinality of number. This may be disguised in the early number operations children carry out because they fall back on counting to find an answer. To achieve an understanding of cardinality, they need considerable experience in associating 'number names' with sets of different objects and arranged in different ways. The name has to be given meaning that is associated with a 'quantity' and not just as a 'count'. The use of structured learning materials has been called into question in this respect (as well as in some other areas of mathematical learning). Where counting and cardinality are concerned, the learning experiences they have had need to be generalizable and they have to come to recognize the applicability of particular numbers to given quantities in a variety of situations. Using structured materials creates an unnecessarily unreal situation and in some cases, they may even disguise the mathematics they are intended to teach.

The language involved in teaching number and working with number generally can also come between the child and the mathematics. The importance of discussion and verbalizing mathematical ideas is emphasized again.

4. *The four operations*

It is evident that children rely on counting to solve much of the mathematics they are given to do long after they have been taught the different operations. Often it appears that the procedures they are taught conflict with children's own strategies and this emphasizes again how knowing these strategies can provide insights into how children think and what their mathematical idiosyncrasies are. Where children may not understand a particular operation and not have a successful model of their own to make the desired progress, they can be guided to adopt more effective models.

As well as a variety of models from which to choose, they benefit from having flexibility in their choice of strategies to be used. The evidence suggests that they will continue to use a strategy that they understand and that works for them. Where algorithms are imposed at too early a stage, they may lose their confidence, and the understanding they have may be undermined. Children's confusion caused by the horizontal and vertical presentation of the same addition task provides an example of this. It may be that whereas they can understand how to break a number down (e.g. $45 = 40 + 5$) they may not understand that 45 is made up of four tens and five units. Knowledge of this kind can either make clear where children's strengths lie or help to pinpoint where their strategies let them down.

Fractions

1. *The nature of fractions*
The evidence indicates that fractions remain one of the most difficult areas of mathematical teaching and learning. It is suggested that this is because fractions do not occur in a 'natural' way in children's everyday lives. The notion of 'equal sharing' for example, is only one of the many 'personalities' fractions have, and in itself is not adequate to convey the concept of fractions to children in a meaningful way (although it is one of the many roads that can lead to an understanding of fractions). It is suggested that, for the teaching of fractions to be effective, they be approached as they arise in realistic situations that demand sharing as well as aspects of measure, ratio and proportion.

2. *Unit fractions and the use of manipulatives*
Two pitfalls identified in the early teaching of fractions appear to be an over-emphasis on unit fractions ($\frac{1}{2}$, $\frac{1}{3}$, $\frac{1}{4}$, etc.) and the use of manipulatives in teaching. These factors could be seen to come together since the structured materials used in the teaching of the concept usually appear as unit fractions. The case is made quite strongly in the research evidence that when the mathematical characteristics are very specifically built into manipulatives, the essence of the mathematical meaning they are meant to represent is lost. A square or rectangular or circular disc divided neatly into halves, thirds or quarters does not give children experience of what, for example, a half is. It is possible that when they place two halves together it may not make sense, as, for example, if they were to put a half of a circular disc and a half a rectangular card together and think they have a 'whole' of something.

3. *Continuous and discrete quantities*
The strands in the teaching of fractions that touch upon equal parts of one whole, equal parts of a continuous quantity and equal parts of a 'whole' that is not one (e.g. an entity that is a $\frac{3}{4}$ piece of something else such as a metre or a pie) need specifically to be developed throughout the teaching of fractions.

4. *Confusion with whole numbers*
Children appear to continue to view fractions as consisting of separate whole numbers, e.g. $\frac{3}{5}$ is seen as *three* and *five* but not as a single object or expression that signifies something called 'three-fifths'. This is especially the case when carrying out operations with fractions and suggests that the notion of a fraction as a mathematical object (in itself) needs to be addressed throughout the teaching.

Measurement

1. *Development of concepts in measurement*
Studies carried out to detect the development of learning in measurement appear not to have identified clear sequential levels. Although children may

appear to have a reasonable grasp of what measurement is about they may, for example, not fully understand the difference between area and perimeter well into their secondary years. It appears to be important that teaching related to measurement at both primary and secondary level needs to be recursive, and return frequently to previous learning to consolidate it in new situations.

2. *Relationships of units*
The relationship between units such as millimetres and centimetres or grams and kilograms has continually to be reviewed. Again, they appear to be best understood in real situations where children have to make judgements about 'heaviest' or 'longest' or 'most' and actually use the units. Doing mechanical number work involving the combining of units is meaningless if the relationships between the measures is not known and understood.

3. *Standard and non-standard units*
There are some indications that using non-standard units to establish the need for standard units can confuse children and may be an unnecessary step in the teaching of measurement generally. The evidence points to the fact that they are not part of the 'real' world and impose a degree of complexity on the learning process that need not be there.

4. *Length and area*
Confusion between linear measurement and the measurement of space persists well into later years of schooling. Children have difficulty with the notion of quantifying space and confuse its measurement with the outlines that define the space. Some of this confusion may be reinforced by the teaching methods used in establishing the concepts.

5. *Materials used in teaching area*
Covering shapes and surfaces with tiles and counting them as an initial way to calculate area can become fixed in children's minds and the tiles as units may, in a sense, become distractors causing children to think of area in terms of counting. They need activities specifically to help them make the leap from using a tiling unit and thinking in counting terms. One strategy that appears to be effective is for children to move to drawing rough grids over shapes and gradually 'fading' these so that the ends or 'points' of markings are left on the lines, defining the shape.

Chapter 2

Shape and Space

INTRODUCTION

Children's earliest mathematical experience is spatial in nature as they physically explore the space around them by moving within it and discovering their relation to it. Some of the earliest concepts built up are concerned with size as they begin to recognize three-dimensional objects (including people) as bigger and smaller, and gradually develop these concepts from a recognition of the *physical occupation of space* to a recognition of *position in space*. These early features of the development of children's spatial awareness underline yet again the importance of activity in early mathematical learning. We have seen (Chapter 1) that early number concepts are formed by internalizing initially one-to-one matching in counting activities of a variety of kinds that involve physical action and touching objects as they count. In a similar way, the establishment of spatial concepts is built on the actions of a child. This development of spatial awareness continues with a two-dimensional representation of the three-dimensional world and further abstraction of the concrete situations they experience and perceive. With time and appropriate experience, spatial awareness eventually develops into spatial reasoning.

Clements and Battista (1992) describe spatial reasoning as consisting of 'the set of cognitive processes by which mental representations for spatial objects, relationships, and transformations are constructed and manipulated' (p. 420). As well as the term 'spatial reasoning', the terms 'spatial ability' and 'spatial visualization' also appear in the literature. With respect to spatial ability, Leder (1992) suggests that 'the lack of definitional precision with respect to both spatial abilities and mathematics has been used to explain an inconsistency in the literature on the expected relationship between spatial ability, mental imagery and mathematics performance' (p. 613). To add to the confusion, Orton (1992) suggests 'it might not be helpful to assume that spatial ability and ability to visualize are the same thing' (p. 119). Whatever the terminology used, it is generally accepted that there is a relationship between cognitive variables of a spatial nature and learning related to geometrical concepts, and there are clear indications that nurturing spatial abilities in children is important to this aspect of their learning. What is less clear is the relationship between spatial ability and other areas of mathematical learning, as we shall see.

In this chapter, we shall begin with a brief consideration of spatial ability and what the literature says about the role this construct plays in the development of children's geometric thinking, including the relationship between spatial skills and gender. Current theoretical perspectives related to the teaching and learning of shape and space topics will be considered next in which the roles of technology and manipulatives will be touched upon. This will be followed by an examination of studies relating to classroom teaching and learning, first in connection with younger children and then with pupils at secondary level. The chapter ends with implications for teaching geometric topics in schools.

SPATIAL ABILITY

Although research may not confirm that spatial ability is necessarily the most vital component of mathematical ability (Orton 1992), it is certainly the case that it plays a vital part in the development of geometric concepts and their representation. There is a considerable body of research evidence that indicates a relationship between spatial ability and achievement in geometrical learning (e.g. Lean and Clements 1981, Johnson and Meade 1987, Ben-Chaim *et al.* 1988). What seems to be in question is the strength of such a link with non-geometrical learning in mathematics. Fennema and Tartre (1985) have found that the relationship between spatial ability and other aspects of mathematical learning is not clear-cut, while Lean and Clements (1981) conclude in their study that 'students who process mathematical information by verbo-logical means outperform students who process their information visually' (p. 144). The fact that one child appears to have a greater level of spatial ability than another does not necessarily mean that the child will achieve greater success in mathematics generally. It would seem the opposite may sometimes be the case and it is important to recognize that other factors play a part as well. Some research suggests that it is possible for spatial awareness to come between the child and other types of learning. Clements and Battista (1992) note that visual learning can sometimes be constrained by the verbal and propositional knowledge children may have, and Hershkowitz's (1989) findings suggest that the use of imagery in mathematical thinking can sometimes cause difficulty, possibly due to the dominant pattern in a child's way of thinking. Research into the functioning of the brain has shown that the right hemisphere of the brain is associated with spatial processes of an artistic and visual nature while the left hemisphere is concerned with processes of a logical and analytic kind (Springer and Deutsch 1981) and the implication is that this could affect an individual's performance in the different kinds of skills being learned. The evidence indicates that, one way or another, either by contributing to or inhibiting the devleopment of mathematical thinking, spatial ability is a major factor in children's learning of geometric skills and concepts.

There is some evidence that spatial ability can be nurtured through appropriate classroom activities (Ben-Chaim *et al.* 1988) which in turn can have a positive effect on children's spatial learning. Within a classroom situation, clearly an appropriate balance is needed of different kinds of activity to cater for the varying degrees of this potential that children may have, and to accommodate their learning patterns. We need to provide children with what Davis (1986) calls 'cognitive building blocks' by creating situations to help them to build up the mental representations related to the study of shape and space. He says, 'Notice, in particular, how mundane and "ordinary" the key experiences are: moving small objects, rotating them, rearranging them into patterns. Many powerful and abstract ideas have origins of this sort' (*ibid.*, p. 279). Actions of this type are essential for children to develop their spatial ability and their ability to reason spatially.

Gender

Geometry is an aspect of mathematics, together with measurement, which has been a focal point for the differences in the mathematical achievement between girls and boys. This has hinged on the research evidence that indicates that girls have a less strongly developed spatial ability compared with boys (Fennema and Tartre 1985, Kaur 1990, Kamira 1992, Geary 1996). The effects of biological differences on mathematical achievements have been studied by Geary (1995) who concluded that 'the cross-national pattern of sex differences in mathematical abilities is generally consistent with the argument that these differences emerge primarily for geometry and other areas where the use of spatial representations might facilitate performance' (p. 247). Battista (1990) found that the single most important factor for predicting girls' performance in geometry was spatial visualization whereas with boys it was a combination of:

(a) spatial visualization; and
(b) the discrepancy between spatial visualization and logical reasoning.

He found that high school boys tend to use spatialization skills when solving geometric tasks which involved a combination of visualizing and thinking it through but not drawing, while girls use drawings to help them towards a solution. The suggestion is that perhaps because boys have a higher degree of spatial ability, they will visualize a solution to a problem (i.e. solve it mentally) rather than resorting to drawing, while girls tend to make drawings to help them reach a solution. This evidence of the role of logical reasoning together with spatial ability suggests that girls and boys use different strategies to solve problems of a spatial nature. There is, however, some conflicting evidence here. Another study (Tartre 1990) found that there was no difference in mathematics achievement (on geometrical and non-geometrical tasks) nor in spatial ability between high school girls and boys. However, girls with lower spatial ability scored lower than boys or girls with higher spatial orientation, indicating that lack of spatial ability appeared to contribute to low achievement.

Kimura (1992) states that 'Major sex difference in intellectual function seems to lie in patterns of ability rather than overall level of intelligence (IQ)' (p. 81). Spatial ability is only one factor contributing to the pattern, and while in girls and boys it appears to be a factor in mathematical performance, Clements and Battista (1992) point out that 'It is important to keep in mind the complexity of the research on gender differences. Such differences are observed in some areas of mathematics but not others, on some spatial tests but not others, in some cultures but not others' (pp. 456–7). However, when writing of results comparing girls' and boys' achievement in mathematics in Singapore, Burton (1990) states that 'results are consistent with those in other societies in that they demonstrate males outperforming females in general, and in particular in quantification and spatial visualization' (Burton 1990, p. 6).

Recent research into the mathematical achievement of girls and boys in the UK has shown that the gap in achievement (in favour of boys) that once existed between them has now been closed (Patrick 1990, Johnson 1996, Elwood and Gipps 1998). This has been widely interpreted as the result of a new awareness of potential gender bias in teaching materials, methodology and forms of assessment, some of which have taken note of the possibilities for such bias. Positive action has been taken (e.g. Burton 1986) and the importance of the effect of the role of teacher expectations with respect to girls' mathematical achievment (Nickson and Prestage 1994, Haastrup and Lindenskov 1994) appears to have been more generally recognized. The results over recent years in Denmark where 'the trend over the past twenty years has proved beneficial to girls' have been acknowledged and this has been at least in part due to 'more geometry, which requires looking for patterns, etc. from the beginning' (Haastrup and Lindenskov 1994, p. 37).

THEORETICAL BACKGROUND

The major theoretical perspectives that influence current research in the teaching and learning of the mathematics of shape and space are essentially cognitivist in nature. They range from a view of the development of geometric thinking as a progression through levels to modelling in geometric problem-solving. More recently, there have been attempts to incorporate the effects of the use of technology in this area of learning into theoretical considerations. We shall begin this section with the van Hiele (1986) theory of levels of geometric thinking. Although this theory is not unchallenged, nor does it inform all of the research reported here, the levels identified are a reminder of the nature of the elements that go to make up geometric thought. As Clements and Battista (1992) state: 'Generally, empirical research from both the US and abroad has confirmed that the van Hiele levels are useful in describing students' geometric concept development, from elementary school to college' (*ibid.,* p. 428).

The van Hiele levels of geometric thinking

The van Hiele theory of development in geometric thinking is described in the following way: 'students progress through levels of thought in geometry from a Gestalt-like visual level through increasingly sophisticated levels of description, analysis, abstraction, and proof' (*ibid.,* p. 426). The four main characteristics of the theory are summarized as follows:

1. Learning is a discontinuous process which involves discrete and different levels of thinking.
2. Levels are sequential and hierarchical, and to move from one to another, an individual must have mastered much of the learning at the lower levels, a progression which is due more to instruction than maturational factors.
3. Concepts that are intrinsically understood at one level become extrinsic at the next, i.e. concepts the learner may not have known they had are manifested in an overt way.
4. There is a particular language associated with each level involving the use of symbols and the relationship between these, so that language structures are an important factor in progression through the levels.

The first of these points is one which is not generally supported by research evidence (Fuys *et al.* 1988, Gutiérrez *et al.* 1991) and it would seem that, rather than being at discrete levels, children oscillate between adjacent levels and even go back to the first level when faced with a new geometric problem. Because language is seen as an extremely important factor in the van Hiele theory and as language is developed from real-world experience and action, manipulatives also take on considerable importance. It is seen as vital that children arrive at their knowledge of shape by handling a variety of manipulatives. This kind of experience, as we shall see, has been found to help children come to understand the properties of shape that lead to their definition, and help them to formulate their own insights about what happens when particular actions are carried out on a shape (Fuys *et al.* 1988, Pallascho *et al.* 1993). The lack of it, on the other hand, can hold children back from progressing in their geometric thinking (Anghileri and Baron 1997).

Schoenfeld (1986) describes the five van Hiele levels with examples which help to clarify the differences that are seen to characterize children's spatial learning as they move from one level to another.

First level: Gestalt recognition of figures. Students recognize entities such as squares and triangles, but they recognize them as wholes; they do not identify the properties or determining characteristics of those figures.

Second level: Analysis of individual figures. Students are capable of defining objects by their properties (e.g. 'a rhombus is a four-sided figure in which all four sides are the same length') but do not see a relationship between classes of objects (e.g. 'a square is a rhombus with a right angle.')

Third level: Analysis of relations. Students can conclude (for example) that every square is a rhombus, because of the properties of squares (e.g. all squares are equal, and a quadrilateral with four sides is a rhombus.) However, they have not attained ...

Fourth level: Deductive competence, the goal state of 10th grade geometry. If asked to prove, for example, that the inscribed angle subtending a given arc of a circle has a measure half that of the central angle subtending the same arc, the student can do so by producing a series of statements that logically justify the conclusion as a consequence of the 'givens'.

Fifth level: Understanding of axiomatics (rarely achieved). Students appreciate the role of axioms and the role of logic in deductive systems, and recognize that Euclidean geometry is one of a number of possible ways to describe an abstract mathematical universe. (*ibid.*, p. 251)

As this description indicates, the levels move from:

1. the visual,
2. to the descriptive and analytic,
3. to the abstract/relational,
4. to formal deduction and finally,
5. to a level characterized as mathematical and involving rigour. (Clements and Battista 1992)

These five levels have been reduced to three and levels 3, 4 and 5 have been collapsed into a single level (van Hiele 1986). However, its is suggested that 'the need for precision in psychologically oriented models of learning argue for maintaining finer delineations' and as a result, Clements and Battista (1992) have not only addressed the original five levels but added a sixth. This they call *Level 0* which they first describe as *pre-recognition* but later is referred to as *pre-representational* by Clements (see Clements *et al.* 1997 below). The levels have had a considerable influence on the teaching of geometry partially because phases of instruction which have been linked with the levels have been found to be effective in classroom teaching (Usiskin 1982). Tests have also been devised to assess the level at which a child is functioning (Mayberry 1983) and the tests themselves have been used as the focus for research (Lawrie and Pegg 1997).

Constructivism and spatial learning

As with children's development of number concepts, much of the work related to the study of the development of spatial skills is derived from Piagetian thought. Two major themes pervade this part of Piagetian theory and help us to understand the rationale for much of the current teaching of geometric concepts and skills.

> First, representations of space are constructed through the progressive organization of the child's motor and internalized actions, resulting in operational systems. Therefore, the representation of space is not a perceptual 'reading off' of the spatial environment, but is the build-up from prior active manipulation of that environment. (Clements and Battista 1992, p. 422)

The active engagement (both physical and mental) of children in learning mathematics of a spatial nature is stressed once again as we have seen in the context of number. However, spatial skills and concepts can be seen to be even more directly related to children's immediate, personal experience than is the case with number simply because of the fact that they themselves occupy and act upon space. An everyday example such as a child placing an orange on a table illustrates how awareness of geometrical properties can arise in very early years. The table surface is flat and the orange is round. Round objects move differently from other objects such as a toy brick – they roll – and so the orange has to be placed carefully so that it will not move. Through everyday experiences like this and by the physical handling of three-dimensional objects, children gradually come to recognize and ultimately to understand what it is that makes one object different from another, i.e. their properties. This brings us to the second statement about Piaget's theory of space:

> Second, the progression of geometric ideas follows a definite order, and this order is more logical than historical in that initially topological relations (for example, connectedness, enclosure and continuity) are constructed, and later projective (rectilinearity) and Euclidean (angularity, parallelism and distance) relations. This has been termed the topological supremacy thesis. (*ibid.*)

As children progress through this development of their ideas from direct experience, they begin to identify the characteristics of the things they handle and what happens to them when they move, and so they build up a repertoire of words that is the language of shape and space. In the Piagetian view, then, when children reach the stage of representing space they are not doing something as simple as reflecting directly what they see in their environment, as already noted. They build up a store of thoughts related to the actions they perform with objects and the language associated with them. The global co-ordination of different visual and tactile experience results in the development of an understanding of projective space (the relationship between the child and what is being explored or acted upon) or three-dimensional space. Gradually, this becomes internalized and synthesized into an awareness of Euclidean or two-dimensional space leading to the development of spatial reasoning.

The constructivist perspective of the learning of spatial concepts and skills which builds upon this Piagetian view of children's development of spatial concepts has much in common with the van Hiele theory of levels of development. Perceptual space is first of all seen to be constructed through the early sensory-motor exploration by the child of the environment (hence manipulatives are as central to this view of spatial development as they are to the progression through the van Hiele levels). Gradually this becomes systematized so that what begins as a tactile experience eventually becomes co-ordinated and abstracted. As Clements and Battista (1992) put it, 'It is as if a space were emptied of objects so as to organize the space itself' (p. 424). The child is no longer thinking of the relationship between themselves and an object but of the relationship between the *position*s of all the objects present within a space. However, they suggest that it may not be quite as straightforward as this and conclude that theory based on topological supremacy is not supported by the evidence: 'Rather, it may be that the ideas of all types develop over time, becoming increasingly integrated and synthesized. These ideas are originally intuitions grounded in action – building, drawing, and perceiving' (p. 426).

There are various criticisms of Piaget's work concerned with children's development of spatial concepts; some, for example, relate to the mathematical accuracy of his use of terminology such as 'topological equivalence' and others relate to the nature of the materials he used with children in his experiments. Clements and Battista (1992) suggest that the results of many of his studies 'may be an artifact of the particular shapes chosen and the ability of young children to identify and name these shapes' (p. 425). However, as with the case of other areas of mathematics, Piagetian studies have provided a foundation and starting point for much of current research into the development of children's spatial thinking and have provided the basis for many of the teaching strategies used in the early years of children's spatial learning.

Cognitive modelling

Different cognitive models exist for the development of spatial skills and learning. They differ from other theoretical interpretations of the development of spatial thinking in the degree of precision they bring to it and attempt 'to integrate research and theoretical work from psychology, philosophy, linguistics, and artificial intelligence' through the construction of models (Clements and Battista 1992, p. 424). Greeno (1986) describes the purpose of models in this way:

When constructing a model reveals that assumptions previously considered sufficient are insufficient, we gain new understanding of the knowledge and processes that are required for student success in the instructional tasks, especially when developing the model reveals gaps we can fill. (*ibid.*, p. 61)

A model, then, may identify important features of a learning situation not previously recognized and arrange these features in such a way as to give a structure to the way particular learning takes place (Schoenfeld 1987) and may in time be adapted to take account of new features in the learning process not previously identified. In this way, they help us to learn something new about how individuals approach solving a problem, including, as we shall see, those of a geometric nature. Thus they help to illuminate how we come to know certain things and how we can intervene to make the process more effective and more efficient. In the process, they set out to account for all the knowledge a learner brings to the learning situation, which includes the explicit knowledge that is identifiable by the learner, and 'unseen' aspects of their learning. Greeno refers to the latter as 'tacit knowledge'. He describes this as 'knowledge needed to perform a task, but whose presence is unsuspected by the performer' where a child may be able to do something but is unable to describe what he or she is doing; the model then becomes 'an explicit hypothesis about both explicit and implicit knowledge' (p. 62). Greeno considers it important that a cognitive model of a particular task addresses both the tacit and the explicit knowledge needed to perform the task. It is as if, by structuring a task in a particular way, it allows the child's tacit knowledge to 'surface' and to contribute to the successful completion of the task.

Greeno (1980) developed a model for problem-solving in geometry which involves the notion of tacit knowledge. In it he identified three types of production reflecting the three domains of geometry that children need in order to solve geometrical problems. These are:

1. *propositions* which are used to make inferences;
2. *perceptual concepts* used to recognize pattern;
3. *strategic principles* used in setting goals and planning.

The first two were included explicitly in teaching materials but the third was not. The model was developed through working with Grade 9 children using think-aloud protocols, and the results ultimately were used to construct a computer simulation to enable the solution of similar types of problem by children. The strategic principles were introduced as the children worked through the problem; these are the 'tacit procedural knowledge' of which they were unaware but were specific to the nature of the problem being solved (i.e. a geometric problem) (Clements and Battista 1992, p. 435). The intention in formulating a model in this way through trialling, observation and analysis was to identify the implicit strategies used by the children and then to incorporate them in a teaching programme to be used with other children (Pallascho *et al.* 1993, Chinnapan 1998a).

On the one hand, this approach may be seen to be contrary to the constructivist perspective because it removes the opportunity for children using the completed programme to evolve their own strategies. The strategies that once were the tacit knowledge of one group are, as it were, being imposed on another. On the basis of this argumentation, any programme designed for teaching could be said to be doing the same thing, the only difference being that in this case there is the declared intention to incorporate the strategies that are identified into a teaching programme. On the other hand, if the strategies were identified for the use of teachers (as is the case in Cognitive Guided Instruction) then it may be seen as helping the teacher to facilitate the children's construction of knowledge. This situation exposes the conundrum of constructivism, i.e. the delicate point at which (and the way in which) intervention is seen to be removing the potential for the child to make his or her own constructions. De Villiers (1998) takes up this point when he refers to an approach called the *reconstructive* approach which 'is characterized not by presenting content as a finished (prefabricated) product, but rather to focus on the genuine mathematical processes by which the content can be developed or reconstructed' (*ibid.*, p. 250). He notes that this 'does not necessarily imply learning by discovery for it may just be a reconstructive explanation' by some other means (*ibid.*).

A major criticism levied against cognitive models of this kind is a lack of explanation about important aspects of children's geometrical learning. Models do not tell us about the process aspects of such learning, for example conjecturing and restructuring and, perhaps more importantly, they do not tell us why a child may not succeed in solving a particular problem (Clements and Battista 1992). It is as though cognitive models provide a simulation of a working structure without giving any insight into 'how' and 'why' of the way it works. However, they do help us to identify where and when the understanding breaks down, and while they may not indicate the reasons, they pinpoint where intervention might best take place to improve understanding. For example, this could happen using the Greeno model where a child's inability to solve a problem might be traced back to the level of propositional knowledge which may be lacking. Similarly, children may have the conceptual knowledge but not have formed the appropriate schema necessary to develop a strategy for approaching a solution (Mitchelmore and White 1998).

Intuition of space

Using his interpretation of intuition, Fischbein (1987) also considers the development of an understanding of space as growing through our physical actions. He sees intuition as being characterized by several properties, amongst them *self-evidence and immediacy* where no extrinsic justification is considered necessary, *perseverance* which gives intuitions their stability, and *extrapolativeness and globality* which means intuitions are seen to offer a unitary, global perspective. He describes the intuition of space as not merely a reflection of the properties of space but as 'a highly complex system of expectations, and progress of action, related to the movements of our body and its parts' (*ibid.*, p. 87). In this view, each individual builds up his or her own theory of space, as it were, as a result of the way in which experiences are stored and become organized. This in turn leads to the formulation of beliefs and expectations with respect to their spatial organization of the world around them. It is argued that one of the weaknesses of this theory is that intuition can give the wrong interpretation of reality since it is not always necessarily correct (Clements and Battista 1992). Intuitions are related to an individual's particular experience within a given cultural setting and are necessarily influenced by this. Hence because our spatial experience is unique to us as individuals and interpreted accordingly, it follows that our interpretation of space may be idiosyncratic insofar as it reflects our personal way of perceiving, experiencing and recording. This means that an individual's spatial awareness can, as a result, be mistaken with respect to the accepted conventions and beliefs in mathematical thinking and activity of this kind.

While Fischbein (1994) acknowledges two other aspects of mathematical behaviour (the formal and the algorithmic) in addition to the intuitive, he considers the intuitive to be predominant to the extent that it can obliterate the formal control and distort or possibly obstruct the appropriate mathematical reaction in a given situation. This is a factor that has clear implications for the early stages of teaching and learning spatial concepts and skills.

The use of manipulatives

There is a considerable literature which supports the positive effects of using manipulatives in teaching spatial concepts (e.g. Fuys *et al.* 1988, Sowell 1989, Gal and Vintner 1997). However, as we noted in connection with number concepts, manipulatives *in themselves* do not necessarily lead to the formation of what are geometrically correct concepts (Gravemeijer 1997, Schoenfeld 1987, Cobb 1987). Research indicates that children tend to limit their concept of geometric shapes to the exemplars that are used in teaching they receive, and when children learn the definition of a shape, whether two-dimensional or three-dimensional, their visual image may remain connected to the prototype they have come to know (Burger and Shaugnessy 1986, Fuys *et al.* 1988, Hershkowitz *et al.* 1990). A child who has only ever handled a cube would have a very limited concept of what a cuboid is, as Gravemeijer points out. Equally, if children do not have the necessary concepts such as edges, sides, corners and faces, definitions that characterize a shape according to its properties will be meaningless (Hasegawa 1997). With the concepts they must also have the language associated with them (Anghileri and Baron 1997). In Fischbein's (1987) view, because concept images are the product of children's intuition, children do not recognize particular cases of a given shape and their images act as general models. Whatever the theoretical perspective, clearly children need to experience a shape in a variety of contexts, both formal and informal, as well as regular and irregular versions of it, so that they come to associate the name of the shape in the different representational forms.

Computer environments

One of the earliest examples of computer technology to be used in classrooms was Logo. Clements and Battista (1992) provide considerable detail of studies carried out with children learning spatial concepts in a Logo environment in their overview of research into geometry and spatial reasoning. While much of this is positive it is important to be aware that some evidence is not. For example, they found (Clements and Battista 1989) that 'Logo experiences may foster some misconceptions of angle measure, including viewing it as the angle of rotation from the vertical' (Clements and Battista 1992, p. 451). They also note that 'students' misconceptions about angle measure and difficulties coordinating the relationships between the turtle's rotation and the constructed angle have persisted for years, especially if not guided by their teachers (Hershkowitz *et al.* 1990, Hoyles and Sutherland 1986, Kieran 1986, Kieran *et al.* 1986)' (*ibid.*). Another problem is that children are not necessarily engaged to think mathematically in the Logo environment and can become too dependent on visual cues (Hillel and Kieran 1988). The importance is stressed of the careful planning and structuring of tasks by teachers when using a Logo environment in their teaching.

Writing of the role of technology in mathematics education, Kaput (1992) notes that '*Continuous transition* of intermediate states is likely to be a cognitively important feature of dynamic systems' (*ibid.*, p. 526, author's italics). He uses the example of Cabri geometry which allows each stage of a transformation to be visible to the learner and thus conveys a stronger sense of the dynamic aspect of geometry and shape. He also describes the 'dragging' action which is built into the system in the following way:

> The movement occurs according to whether and how a given part of the construction is logically constrained; an unconstrained vertex can be moved freely, whereas those points and lines which depend on that vertex must follow accordingly, helping to reveal underlying invariance in a new way. For example if one constructs the three medians of a triangle (and notes that they intersect in a point), then one can drag any vertex of a triangle, thereby changing its shape, but the medians remain medians and they continue to intersect in a single point. (*ibid.*, p. 538)

Other systems such as Supposer and GeoDraw have variations of a similar action but, whereas Supposer includes non-Euclidean axioms, Cabri supports Euclidean transformation geometry in a dynamic way.

There is an ever-increasing literature on the use of technology in geometry in particular, within the mathematics curriculum (e.g., Laborde and Laborde

1995, Noss and Hoyles 1996, Clements *et al.* 1996, Jones 1997, Gardiner and Hudson 1998). At least one theory (Lins 1998) is emerging which attempts to take into account the effects of a single computer environment (Cabri geometry) and others relate to computer environments more generally (e.g. Noss and Hoyles 1996).

THE DEVELOPMENT OF SPATIAL ABILITIES IN YOUNG CHILDREN

The theoretical backgrounds we have just considered indicate the crucial nature of children's early learning about shape and space. Research related to the teaching and learning of spatial skills and concepts in the early years focuses on a wide range of topics, from looking at how children perceive shape (Clements *et al.* 1996) to more specific matters such as the concept of angle (Mitchelmore and White 1998). The studies reported here provide examples of current theoretical perspectives and methodologies used in the study of children's spatial leaning.

Early perceptions of shape

Young children's early conceptions of geometric shapes were studied by Clements *et al.* (1997) with a view to producing a detailed description of what these ideas are. They set out to identify:

(a) what criteria pre-school children use to distinguish one shape from another;
(b) whether they use these criteria consistently; and
(c) whether the content, complexity and stability of these criteria are related to either age or gender.

Individual clinical interviews were carried out with 97 children aged from 3.5 to 6.9 (pre-school in the USA) during which they performed shape selection tasks using paper and pencil. The shapes used were the circle, square, triangle and rectangle and the children had to mark a given shape which appeared on a sheet along with visual distractors. The final task was more complicated and involved overlapping circles and squares together from which they had to identify circles. Responses were coded as 'visual' (where children had to look at the whole shape) or 'property' (where they used properties of shape in making their choice). The 6-year-olds performed considerably better than the younger children, and they found that children identified circles with a high degree of accuracy, most of them describing circles visually with no mention of any property. They were slightly less accurate in identifying squares, and occasionally properties were used when giving reasons for their selection. Children were not as accurate in identifying triangles and rectangles with only a few responses in which properties were mentioned. The 4-year-olds tended to accept squares as rectangles, and properties were referred to less often for rectangles than triangles or squares. A significant difference was shown across age groups in the final complex configuration involving squares and circles, where 6-year-olds were most successful. No gender differences are reported.

The researchers argue from these results that 'children who reliably distinguish circles or triangles from squares should be classified as pre-representational' and not at the visual level of van Hiele's Level 1 (Clements *et al.*, p. 165). Children at this pre-representational level (Level 0) are just beginning to form schema as a result of the unconscious linking of relationships between geometric figures with specific patterns. Second, they argue for a reconceptualization of the van Hiele Level 1. Their results suggest that at this level children are not merely interpreting shapes visually but they are thinking in a more syncretic way and bringing together visual responses with some recognition of the components and properties of shape. Even though children may be expressing reasons why a shape should not be in a particular class, they are doing so because of a degree of recognition of 'difference' despite the fact that they cannot give it a name. As Clements *et al.* (1997) say 'the contrast between the figure and the prototype provokes descriptions of difference' (p. 166). However, the children have not yet had sufficient experience nor developed the language in association with a shape to identify the differences and give them their name. But, as the children progressed from one level to another, they paid more attention to the components and properties of shape and, as a result, they made more mistakes. An example given was when, having described a square as a shape with 'four sides equal and four points', they also classified a rhombus as a square on the basis of this level of knowledge of the properties of the shape (*idem*). They were not merely identifying four angles and four sides but recognized the sides as being equal. What they had not yet recognized was the notion of equality in relation to angles and this led them to classify the rhombus incorrectly. This study exemplifies the kind of progression children move through when they begin to synthesize their perceptions of given shapes according to the properties they have come to recognize. At this stage, the children in this sample have yet to develop fully the concept of angle as a property that contributes to the definition of a shape as in the case of the square and rhombus.

Hasegawa (1997) also noted children's difficulties in focusing on the properties of shape and their inability in relation to tasks (e.g. rotation) with basic shapes. He notes that 'The definition of a quadrilateral for example, contains the word "line", but the term cannot be defined mathematically on an elementary level' (*ibid.*, p. 176). In identifying shapes according to their properties in the early years of teaching, children are being drawn to using concepts of an abstract kind and of which they have little understanding (as in the case of angle in the study referred to above). Identifying a line is a conceptually difficult task and directing their attention to a line as such is not a simple matter. Hasegawa (1997) studied Grade 2 children's concept formation of quadrilaterals and triangles and in it he identifies four ways to judge a polygon (what he calls an *n-gon*) correctly:

1. counting the number of sides;
2. counting the number of its vertices;
3. using a motion (a congruent transformation);
4. using a general transformation. (*ibid.*, p. 159)

He suggests that 'We can design various situations in which mathematical activities contribute to the constitution of mental objects rather than the formation of concepts' (*idem*). The intention is to make clear the 'object' that is a line. One of the strategies he used to draw children's attention to the properties of the lines that make up two-dimensional shapes was to ask them to draw cages around animals with a view to focusing their attention on the rectilinearity of the edges and corners. Asking the children to colour the inside of cages helped to emphasize this focus more strongly. Even if we choose to describe the properties of a shape in terms of *sides* rather than *lines*, the level of abstraction is the same. The expectation is that the children will focus on what *we* know is a side or line when this often may not be what happens; it is difficult to focus on something that is not recognized.

Levels of spatial thinking in younger children

The van Hiele levels were studied by Fuys *et al.* (1988) when they explored the geometrical thinking of Grade 6 and Grade 9 children (approximately 11 and 15 years old respectively) using a working model of the levels. The children's ability to progress as they were being taught, through and between the different levels, was monitored using assessment interviews. They found that some of the children moved flexibly to different levels while others remained on a plateau. There was also evidence that children moved backwards and forwards between levels and, when studying a new concept, they often lapsed to level 1 thinking but quickly moved to the higher level they had achieved in concepts they had already learned. These findings are similar to those of Gutiérrez *et al.* (1991) who undertook a study in which they focused on pupils' capacity to work at the different levels rather than categorizing them as being at a particular level. Using a vector component system where each level was seen as a component, they tried to determine where on each level a child might be functioning and found that some children were developing at two levels simultaneously; having identified a child at level N, they found components of N + 1 in the child's thinking, thus indicating that some children were working at two levels at once. They argue that children in this situation were seen to be functioning at N + 1 without having firmly consolidated understanding and performance at N, possibly as a result of the teaching they had experienced.

The concept of angle

A study by Mitchelmore and White (1998) set out to explore how well children recognize angle in everyday life. Essentially the focus was an exploration of children's connections between the idea of angle as the mathematical idea of 'amount of turn' and their everyday experience of angles in the physical environment. The work was carried out with a gender-balanced sample of 144 children in Grades 2, 4 and 6 in schools in Sydney. The situations selected for use in the study included nine examples of physical angles in the environment: wheel, door, scissors, fan, signpost, hill, road junction, tile and walls, which were considered to represent a reasonable balance of dynamic and static examples of angle. Situations were devised which focused on each of these examples. The 'fixed situations' were presented in three models: 'neutral (0–90°)', an angle of 45° and a 'middle angle' (22.5° or 67.5°). These examples of angle were presented to the children in tasks involving three different pairings of type of model: moveable–moveable, fixed–moveable and fixed–fixed. In one task, children were asked whether they could identify similarities between the two situations presented in each pair and their responses were classified as dynamic (where they referred to movement) or static (where they referred to a common geometrical configuration). In the second task children were asked to 'do the same as this' by duplicating the interviewer's action in setting a model at an angle of 45° and in the third task they were asked to bend a straw to indicate how they knew both were the same.

The results showed that children across all three grades almost never saw a dynamic similarity between any fixed situation and another, whereas static similarities were readily seen between fixed–moveable and fixed–fixed situations (this recognition increased from Grade 2 to 6). Success rates in the second task (i.e. setting a model at 45°) showed a similarity in response across the grades. Statistical analysis of the data together with observational data suggested that a common construct was being measured by each of the three tasks. The analysis also showed one main cluster of the situations within each grade which indicated that children saw similarities in these situations. In Grade 2 the cluster was – *walls, junction and tile*; in Grade 4 the main cluster expanded to include *walls, junction, tile and scissors*, with two secondary clusters – door and fan, hill and signpost; in Grade 6 the single main cluster comprised *walls, junction, tiles, scissors, signpost and fan*. Thus there was a core of situations perceived as similar by the children, which expanded from Grade 2 to Grade 6. The results of the analysis were presented in the facet model below (Figure 2.1), which shows the basic core of static angle situations progressing outwards to the concepts of wheel in the outer space. This suggests that the stages in teaching the concept of angle using real-world examples should begin with static situations such as wall, junction or tile, moving on to scissors, then fan and signpost, next door and hill and, finally wheel.

These examples move from those with clearly visible arms to those where the arms have to be imagined (as in the case of the wheel). Teaching according to the sequencing in the model would incorporate situations involving this

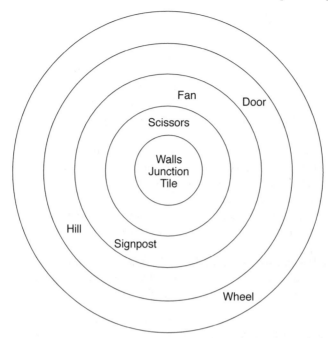

Figure 2.1: Modulating facet model of static similarity between nine situations. (Mitchelmore and White 1998, p. 276)

kind of progression, and angle comparison could be seen as 'amount of opening' from one situation to another. In any situation where the 'lines' (i.e. the rays) are missing, the aim would be to produce these so as to make the idea of angle explicit.

The findings of this study have important implications for the way we teach angle and for our preconceptions about how the concept of angle is related to the real-world situation of the child. As the authors say, the results 'call into question the received wisdom of defining angle as an amount of turning' (Mitchelmore and White 1998, p. 277). It seems that children at these earlier stages:

(a) do not see all movement in angle situations as being the same kind of movement; and
(b) they do not associate angle in a static situation as having anything to do with movement, and these perceptions do not change across the age group of the sample.

They state that 'The "static similarity" spans both moveable and fixed situations; they all involve two lines meeting at a point' which is interpreted as indicating that 'there is no common attribute relating the two lines in different situations. For example, the angle between the two lines *may* represent an amount of turning but it depends on recognizing that all these attributes are similar' (*idem,* authors' emphasis).

The space between the two lines or rays forming an angle may be interpreted in a variety of ways, for example in terms of steepness of the angle of a hill or openness of scissors. The conclusion is drawn that the main feature for children learning the concept of angle is the number of visible arms forming the angle and that '"dynamic similarity" is not a sound basis for developing a general angle concept' (*ibid.*, p. 277). They suggest that the teaching of angle should begin with the central core of their model and move outwards:

> A first idea of angle could be easily established, even as early as Grade 2, from situations with two clearly visible arms – provided examples are not restricted to right-angled corners. The next step would be to seek similarities between these situations and fixed and moveable situations where both arms of the angle are present but not so easily identified (the next two partitions of our model). (*idem*)

They advocate that size of angle should be dealt with without any reference to turning but rather in terms of something like 'amount of opening' to identify the similarity amongst all the situations.

Angle in everyday situations

Another study investigating children's early concepts of angle is reported by Magina and Hoyles (1997) which once again attempted to relate young children's perceptions in the context of an everyday situation, in this case measuring time with an analogue watch. The work was carried out with 54 children from three age groups (ranging from 6 to 14 years old) in a middle-class school in Brazil. The sample was chosen because it could be assumed that these children had been taught the same curriculum and had experienced fairly consistent teaching patterns. In the review of the literature, children's misconceptions in relation to angle are noted (e.g. APU 1980, Hershkowitz *et al.* 1990) which include an inability to recognize the acute, obtuse and right angle in other than a standard vertical or horizontal presentation, and the confusion children have between size of an angle and the length of its rays.

Cardboard watches, some round and some oval in shape and of different sizes, were used in the investigation. The children carried out five activities in which they were asked first of all:

(a) to predict the position of the minute hand half an hour before the time indicated on the watch; then
(b) to move the minute hand to the position it should have after half an hour and to justify their conclusion.

They were also asked to 'compare' lengths of time shown, although all times were a half an hour. The strategies used by the children when making their predictions were analysed and four groups emerged from the analysis of this stage:

1. *no strategy* where children either gave no explanation or their prediction was irrelevant;
2. *fixed representation of 'half'* where children consistently associated half an hour with the number 6;
3. *moving across numbers* where intervals were being described by the children in terms of the hand 'moving across' a number of spaces, although some showed uncertainty about how this should be carried out in all of the activities;
4. *co-ordinated strategy* where children were able to give a variety of ideas related to their explanations of 'half an hour'.

The results of this part of the study suggested that the first step in coming to recognize half an hour is the connection between half an hour and a specific number (6) and has nothing to do with the idea of turning or space.

The outcomes of the activities were analysed and the children divided into three groups accordingly.

- *Group 1.* This group included 6- to 7-year-olds who showed no obvious strategy when they were asked to predict the position of the minute hand after half an hour and also had difficulty turning the hand, whether asked to turn it a 'half turn' or 'half an hour'. In comparing lengths of time these children judged the time according to 'the figure formed by the hands of the watch' (e.g. when the hands ended up at 2:45 they saw this as the largest angle and judged it to be measuring the longer time (*ibid.*, p. 112).
- *Group 2.* The second group was made up mostly of 8- to 11-year-olds. They had difficulty coping with comparisons but were better than the first group when predicting where the hand would be after half an hour, although they tended to be more successful in these tasks when the minute hand was at 6. It was concluded that these children could read the time on a watch well but could not co-ordinate this with measuring intervals of time between one setting and another.
- *Group 3.* The 12- to 14-year-olds performed well in both the prediction and performance tasks and were able to co-ordinate the spatial position of the hands when reading the time and measuring time intervals. They understood the role '6' plays when dealing with half-hours and could read half-hour intervals successfully whatever the shape or size of the watch.

The general conclusion drawn from this investigation of children's understanding of rotation and angle is that 'it appears that children start out by considering the figural properties of space as fixed and only later come to view them as related to the transformations from which they result' (*ibid.*, p. 114) confirming Mitchelmore and White's findings above.

Angle in a Logo environment

An investigation by Clements *et al.* (1996) in which children used Turtle Paths set out to explore children's understanding of the *amount of turn* in work with angles. This study built on previous research evidence (Clements *et al.* 1990) that had established the following: 'After working in Logo contexts designed to address ideas of angle and turn, children develop mathematically correct, coherent, and abstract ideas about these concepts (Clements and Battista 1990, Kieran 1986, Noss 1987)' (*ibid.,* p. 314). A case study approach was used with two girls and two boys in Grade 3, after which further evidence was gathered from work with two classes. The ideas being studied were:

(a) paths and lengths of paths;
(b) turns in paths; and
(c) paths with the same lengths involving isometric exercises.

They found that two overarching themes pervaded the study, the first being perhaps the obvious one of the role of the computer environment in the development of children's concept of turn. The second, however, was more complicated. They found a dialectical relationship between 'two cognitive schemes, extrinsic perspective and intrinsic perspective, in students' knowledge of turns' (*ibid.,* p. 318). They describe the *extrinsic perspective* in terms of a frame of reference that is fixed and imposed from the outside, and the example given of this kind of perspective is co-ordinate geometry. The turn is determined by its relationship to something fixed and external to it. The *internal perspective*, on the other hand, is that of the turtle and here, of course, the children are in control and can direct the turtle without any external constraint. The turns can be made wherever and whenever they want.

  The results of the study showed four more specific themes emerging which were related to procedural knowledge:

1. the concept of turn;
2. right and left directionality (clockwise and anticlockwise);
3. the measure of turn; and
4. combining turns.

The researchers concluded that 'Turns are less salient for children than motion from one position to another' (Clements *et al.* 1996, p. 332). They do not 'leave a trace' and as a result, it is difficult for children to distinguish angles in and for themselves. The example is given of a bicycle which is perceived by children as moving forward in a linear motion which they do not connect in any way with the notion of angle or turn. This emphasizes yet again that what may seem to us to be obvious 'real-world' contexts in mathematics teaching (particularly those that are geometric in nature) are not necessarily obvious to them. The reason suggested for children's difficulties is the fact that a turn is represented by two straight-line segments being joined at a point, and angle has something to do with the space between these line segments which, as noted, is a sophisticated concept. The conclusion is drawn that there are three aspects to the conceptualization of a turn:

1.   Children have to maintain a record of mental images of the initial and final heading of an object as it turns, using an external frame of reference in order to fix the heading.
2.   Children have to 're-present' the activity of the rotation of an object from its initial to its final heading and compare these images to one or more iterations of the internalized image as it moves through a unit turn (30° or 90° in the study).
3.   Having to deal with a multiplicity of ideas at one time adds to the difficulty children have in learning about turn measurement, i.e. they are having to deal with both length and turn and the relationship between the two.

These points are reminders of the complexity of the concept of angle and of the importance of what was noted earlier with respect to Logo, i.e. that the misconceptions formed by children about angle in a Logo environment can be long lasting and difficult to overcome. In spite of these difficulties, children involved in the study were considered to have made 'valuable cognitive constructions concerning rotation' in the course of the investigation' (*ibid.*, p. 334).

Children's concepts of three-dimensional shape

The development of children's ideas of the relationships between three-dimensional shapes have been investigated by Anghilieri and Baron (1997). Poleidoblocs were used in the study with 66 children in reception (4-plus years) and first year (5-plus years) classes in three schools, and seven individual children aged between 21 months and 7 years old. Poleidoblocs are described as '[coloured] bricks in six basic shapes: cubes, cuboids, cylinders, triangular prisms, cones and pyramids, that are interrelated in a variety of ways' (*op. cit.*, p. 12). Practical activities were chosen in order to minimize differences in the development of mathematical language by the children. Some of the activities involved free play and others were structured. Videos were made of 41 of the children during free play using the blocks and all 66 children were videoed as they undertook five practical tasks. During free-play activities, five kinds of mathematical experience were identified:

1.   *sorting* using shape, colour, size and function;
2.   *balancing* and intuitive notions of measurement;
3.   ideas related to *stability*;
4.   aspects of *parallel, sloping* and *horizontal faces*, together with *central positioning*;
5.   the concept of *symmetry*.

These arose from the three categories of structure the children made which included a single tower, a symmetrical structure or structures classified as complex. There were some differences noted between the girls' and boys' activities during the free-play activities, for example while the boys made tall and carefully balanced structures, the girls seemed to focus on sorting, and while boys built trains, girls built structures like a playground and a palace. Colour also played more part in the girls' sorting and building activities.

The different kinds of mathematical experience identified during free play formed the basis for the following structured tasks that children were given to do at the beginning of the project:

1. Matching three-dimensional shapes and two-dimensional faces (matching correct shape face to six faces on a sheet of paper).
2. Building a tower beside another tower and matching its height.
3. A tactile task where a cylinder was shown, then the child had to select the same shape from a 'feely bag'.
4. Sorting where all the bricks the same as the green cylinder had to be selected from a set of ten, then all the bricks the same shape.
5. Continuing a sequence of red bricks in the order: cube, cuboid, cylinder, cube, cuboid, and two more bricks had to be placed.

In the initial test, children found most difficulty with some of the matching tasks (e.g. the matching of a triangular face of a prism to a drawing) and the sequencing task. The children were tested again some eight weeks later, during which time about half the sample had further experience using the bricks in a classroom situation. It was found that there was improvement for all children on many of the tasks, with the greatest improvement by those who had access to the blocks. There was some deterioration, however, in all groups with respect to the tactile, sorting and sequencing tasks, all of which involved recognition of characteristics of the shapes in some way. The reason for this varied with the task, e.g. a flat rectangular cuboid was added to the set of cubes because of their strong identification with the square in their sorting procedures. It was noted that 'It was significant that there was no related discussion with any of the cohort whereby language and concepts of the cube and its square faces may have been clarified' thus supporting other evidence (Saad and Davis 1997) of the importance of language in the development of spatial abilities (*ibid.*, p. 17).

The conclusions drawn from the investigation were that children's learning about three-dimensional shapes achieved by handling and using them in constructions can be improved upon through experience with the shapes, where only tactile and visual experiences are involved. However, when classification becomes part of the task, active engagement of this sort is not enough in itself. It is stated that:

Discussion involving the names and characteristics of the 3-D shapes is necessary for children to clarify mathematical understanding, for example, in the relationships between cubes and squares and the classification of cylinders that look different, being tall and thin or short and fat. (*idem*)

The children in this study were clearly at what Clements *et al.* (1997) have called the 'pre-representational' level in the van Hiele hierarchy and if they had developed through discussion the language to deal with the observations they were making in their activities with the blocks, perhaps there would not have been the deterioration in performance that took place on particular tasks.

Two-dimensional arrays

The development of 7- and 8-year-old children's spatial skills was studied by Battista *et al.* (1998) working with two-dimensional arrays of square tiles. This work built on the results of an earlier investigation with Grade 3 children as they carried out spatial structuring tasks. It was found that these children were disorganized in the way they approached counting the edges of triangular prisms they themselves had constructed, and had no global schema in the early stages to count in 'small groups of composites' (*ibid.,* p. 505). An intervention programme was devised to help them develop a schema for organizing their counting strategies in which it was pointed out that it is helpful to count the lateral edges first, and then those on the base. Through this process the children developed units with which to work, where they grouped the edges according to a common criterion (in this case, the lateral or vertical, and the base or the horizontal) before counting them in these 'composites'. This led to the identification of the importance of the mental process of structuring for children's understanding in spatial situations involving quantifying. Spatial structuring is defined as the 'mental operation of constructing an organisation or form for an object or set of objects' (*ibid.,* p. 503). It is seen as a kind of abstraction, and by applying it in spatial situations, children:

(a) have to identify the spatial components of the situation;
(b) combine these spatial components into composites;
(c) establish an inter-relationship between the components and the composites.

It is suggested that children 'progress beyond this stage when they construct the notion of perspective, recognize that the orthogonal views must somehow be coordinated, and become capable of accomplishing such coordination' (*ibid.,* p. 505). In other words, they have to combine two different views of a single physical situation in the course of spatial quantifying and process that information to make sense of it. How they do this was the focus for the second study.

There were two stages in this second study when the sample consisted of twelve 7- and 8-year-olds. During the first stage, the children were interviewed as they undertook tasks to see if there was a pattern in their structuring and counting strategies, and to distinguish between their ways of operating. In the second stage, this data was analysed with the intention of producing descriptions of the mechanisms the children used in their counting and, from these, if possible, to identify levels in the way they approached the counting. They were given square tiles which had to be fitted into a rectangle and were asked:

(a) to predict how many would be needed to completely fill the rectangle;
(b) to draw where they thought the squares would be placed;
(c) then to cover the rectangles with the tiles.

Three levels were identified in the children's approaches to the counting. At the first level there was complete lack of any row-by-column structuring in their counting, and at the second level there was partial structuring. The third level was identified in two stages, the first of which was level 3A where there was evidence of structuring in sets of row-by-column composites; at level 3B a higher degree of abstraction was reached when children were able to carry out the counting visually, iterating either by row or by column, i.e. by using either a row or a column as the unit. This is interpreted as evidence of how children proceed from the local to the global in their structuring of a spatial task and how they move from using haphazard methods and counting in unit cubes in an unsystematic way, to forming larger 'wholes' (rows or columns). They treat these 'wholes' as units, find out how many of these there are and, ultimately, find out how many of the basic units there are (in this case, the unit tile). Battista *et al.* (1998) stress that these levels are not seen in terms of a general developmental scheme but are evidence of how children function when they approach tasks which specifically involve rectangular arrays. They note the clear implications of these findings for other areas of mathematics, in particular multiplication and in the context of measure. Children who have not reached a stage of seeing arrays in composite units will find it difficult to group objects or symbols to represent the multiplicative structure when, for example, they are shown two rows of three apples to depict 2×3. A similar situation exists when they come to calculate area.

Cabri geometry with 8- to 11-year-olds

In a study by Ainley and Pratt (1995a) of 8- to 11-year-old children's spatial learning using laptops and Cabri software, Papert's idea of *constructionism* is used to interpret the children's experiences using computers in this context. Constructionism describes 'learning approaches in which the learner constructs knowledge through building a meaningful product' (*ibid.*, p. 98). The software was chosen because it was judged to provide children with the opportunity 'to explore and build ideas', but at the same time it was noted that 'in doing so, there was the likelihood of them stumbling across mathematical ideas' (*idem*). The kind of problems that might occur were identified in advance, however. There was a concern that because the software deals essentially with Euclidean geometry, it seemed possible it would not be inviting enough for children of this age to engage with. Also, if children are encouraged to explore in this way, the question arose as to whether they would come across powerful ideas such as 'functional dependence', and if so, how would they cope? Two episodes are described in the study which suggest that rather than being intimidated, children engaged in tasks at two levels in a meaningful way. The 8- to 9-year-olds were able to construct a football pitch after being shown some simple techniques. The interchange between researcher and pupil shows how intervention at an appropriate moment is important in leading the children to an understanding of what a 'perpendicular bisector' could do. The older children were able to build on an idea shown them by the researcher of how to build an equilateral triangle from two intersecting circles and turn this into a procedure for generating nesting triangles. They use Wilensky's (1993) idea of concretion to describe what was happening. Children, by doing mathematics, 'enter into an increasingly informed relationship with the mathematical concepts embedded within that activity' and the more connections there are made between a mathematical object and other objects, 'the more concrete it becomes for the learner' (*ibid.,* p. 104). The researchers stress that it is important to emphasize the purpose of an activity of this sort so that it becomes transparent to the children and the new mathematical concepts arrived at can become functional in other mathematical learning situations.

THE DEVELOPMENT OF SPATIAL ABILITIES IN OLDER CHILDREN

Mental models of spatial concepts

A mental modelling approach was used by Chinnappan (1998) as a tool in studying children's problem-solving in geometry. The focus of the research was to examine 'the potential links between mental models constructed by students, the organisational quality of students' prior knowledge, and the use of that knowledge during problem-solving' (p. 202).

Chinnapan notes research (Prawat 1989) that suggests that 'the effective use of prior topic knowledge during problem-solving is dependent upon the organisation of that knowledge' (Chinnapan 1998b, p. 184). When pupils' knowledge base is a 'well organized or integrated domain', it not only facilitates their accessing of that knowledge in a problem-solving situation but also affects how that knowledge is used in the process of finding a solution. The notion of schemas in a mental modelling approach is discussed and, drawing on the work of Marshall (1995), Chinnapan (1998a) describes a schema as: 'a cluster of knowledge that contains information about core concepts, the relations between these concepts and knowledge about when and how to use them' (*ibid.,* p. 202).

These organized knowledge structures are viewed in terms of:

(a) the *acquisition* of mathematical concepts, principles, procedures;
(b) the *organization* of these into schemas;
(c) leading to the *provision* of a knowledge base for further mathematical activity.

He notes that the more elaborate a schema, the more likely pupils will be able both to construct useful, as well as multiple, representations of a problem, and he set out to study the extent to which such schemas may be used by pupils in their organization of geometrical knowledge to solve problems.

The study involved 30 Year 10 pupils in an Australian secondary school: fifteen high achievers and fifteen low achievers. They were asked to solve a plane geometry problem in individual interview situations three weeks after being taught the relevant geometry topic in class. They were presented with two versions of the task, a *problem context* and a *non-problem context*. The problem consisted of a statement and a diagram and the first task was to find the length of a segment in the diagram. This version of the task was the 'problem context' in which the pupils' schemas were to be studied. The second task related to the same diagram but was open-ended and pupils were asked:

(a) to identify all the geometric shapes they could see in the diagram;
(b) to state any rules, theorems or formulae that they knew in connection with any of the shapes.

This was the non-problem context in which their schema would be considered, and in the second phase of the task, they were told they could expand the diagram in any way they wished but had to address (a) and (b) with respect to any new shapes they might produce. Each interview was videoed.

The pupils' performance on these tasks was analysed and the schemas activated in the process of doing the tasks were classified under the two headings of 'problem context' and 'non-problem context'. In the problem context, a frequency count was made of the schemas, and in the non-problem context they were grouped into four sub-categories:

1. schema used in analysing the diagram (open-ended);
2. schema used from (1) in the solution of the problem (problem-relevant);
3. schema activated as a result of extending the diagram (open-ended);
4. schema used from (3) relevant to the solution of the problem (problem-relevant).

It was found that the high achievers activated more than twice as many schemas as the low achievers in the 'problem context' task (108 compared with 52) but each group activated almost equal numbers in the non-problem context (285 compared with 296). The results in the problem context task were as expected and are consistent with the findings of Prawat (1989) and Schoenfeld (1987), and it is suggested that 'The better structured knowledge base of the high achievers of this study appear to drive moves during their solution attempts' (Chinnapan 1998b, p. 189). They could access both specific knowledge schemas and relevant problem-solving schemas to solve the problem.

The results of the non-problem context, however, were not as expected. Here it was found that the amount of geometric knowledge held by each group was almost equal and it is suggested that the key factor in this situation was the absence of a problem goal, i.e. because there was not a set question to be answered, the low achievers felt no constraints in their approach to this task. Thus it would seem that the low achievers have a substantial store of geometric knowledge but are unable to invoke it when required to solve specific problems. They apparently also had difficulty in using the available knowledge they had to build up a model of the problem and make the links between what they already knew with the content of the problem context task. For example, although they may have identified the right-angled triangle in the figure given, they were unable to use this knowledge to find the length of one of the sides.

The general conclusion drawn is that while the high achievers were able to form 'meaningful, integrated mental representations which showed the link between the givens and the problem's goal', the low achievers 'tended to focus on superficial aspects and hence did not see the connections among the important factors of the problem and their own schemas of geometry' (*ibid.*, p. 213). Chinnapan points out that although the study is based on a single activity and needs to be replicated, it does suggest that we need to develop learning situations that help low achievers in particular to develop the necessary knowledge structures that will help them to become more effective problem-solvers. These pupils have not formed effective schema as a result of not recognizing the relationships between the relevant concepts that lead to the solution of the problem. Learning situations need somehow to focus on helping them to establish such relationships to provide structures for more successful problem-solving.

Perceptions of shape angle

Orton (1997) studied the relationship between children's perception of pattern and their general ability, and issues related to manipulating two-dimensional shapes. The research was carried out with nearly 200 children from 9 to 16 years. Pattern was not seen as being confined to repeating patterns but 'included ideas of shape recognition, congruence and symmetry' and thus was linked with transformations (p. 304). She found that children aged 9 to 10 already have a clearly established body of pattern recognition or knowledge but lack the vocabulary with which to describe it, and as a result it proved difficult to determine what mental transformations were made in their identification of the pattern in shapes in the tasks set. The results indicate three developmental stages in pattern recognition:

Stage 1: copying a shape; detection of embedded pictures; simple completion of pattern; matching picture shapes; recognition of reflection in a 'vertical' axis; simple rotation and reflection; completing tasks with a frame of reference.

Stage 2: matching of embedded shapes; matching of simple geometric shapes in different orientations; more complex reflection and rotation tasks with a frame of reference.

Stage 3: matching of more complex shapes in different orientations; more complex completion of pattern tasks including rotation; recognition of most reflection and rotation. (*ibid.,* p. 310)

One of the main results from the study was that although pupils may often perceive pattern, they lack the vocabulary with which to describe what they see, thus reinforcing the importance of language in developing spatial abilities (Saad and Davis 1997, Anghileri and Baron 1997). They also found that a frame of reference was used frequently to simplify transformations, for example in imagining a vertical line through the letter **S** to distinguish the appropriate transformation in a matching exercise. There was a progression in mean scores on the test as a whole from Year 5 (9- to 10-year-olds) to Year 7 (11- to 12-year-olds) and Year 7 to Year 9 (13- to 14-year-olds) but a decrease from Year 9 to Year 11 (15- to 16-year-olds). The suggested reason in the dip in Year 11 scores was that there was a large proportion of less able girls in that group.

In a study which was embedded in the Cognitive Guided Instruction paradigm (see Chapter 1) Gal and Vinner (1997) explored children's difficulties in understanding perpendicularity and teachers' difficulties explaining it. The research involved Grade 9 pupils described as 'slow learners' with pre-service teachers who were videoed during lessons. Pupils and teachers were interviewed after the lessons to explore the difficulties that arose during them, and the difficulties were subsequently analysed to find the cause of lack of understanding or misunderstanding. They identify six components that make up the concept of perpendicularity that either are implicit or explicit:

1. two lines (either two segments or one line and one segment);
2. the line intersects;
3. a right angle is formed at the intersection point;
4. there are three other angles at the intersection point;
5. these three angles are also right angles;
6. a right angle is a 90° angle. (*ibid.*, p. 282)

Three aspects are noted with respect to the difficulties pupils encountered. The first was *identification* where there was a problem identifying a right angle in its basic form where, for example, a pupil would not necessarily see an angle between two intersecting lines and hence would not know to look for a right angle. Second, *selection* was a difficulty where the pupils had to decide which of the angles presented by the intersecting lines might be a right angle (but not assuming that all four were), identified as 'visual understanding' as described by Level 1 in the van Hiele hierarchy. Finally, *inference,* where identifying one angle as a right angle means that the other three angles must also be right angles, and the researchers point out that this is no simple matter. Teachers sometimes identify one angle as a right angle in a lesson and then as they elaborate on the perpendicularity of the intersection, they may point to any one of the other three angles; unless the pupil already knows about perpendicularity, they may not assume that the second angle indicated by the teacher is also a right angle. (This is similar to the situation noted earlier with respect to the 'lines' forming the sides of a shape.)

The results showed that confusion about perpendicularity on the part of pupils arose from a combination of the various components that go to make up perpendicularity as noted above. This could result from:

(a) the way the visual information is processed (e.g., whether they make any connection between what they see in the example given them with some other external situation);
(b) children's thought processes (e.g., whether they have acquired the relevant concepts necessary such as angle and intersection); or
(c) children's concept images (e.g., whether the relationships among the concepts they have formed come together in such a way so that they perceive the new concept of perpendicularity).

They conclude that for understanding to be reached, the 'visual links between right angles, perpendicular lines, and the complex figures in which they appear' need to be specifically addressed and it is suggested that this can be done using concrete models, drawings and mental verifications (*ibid.,* p. 287).

Computer environment studies

Jones (1997) reports a study in connection with the work of 12-year-olds using Cabri geometry. He refers to the work of Wertsch (1991) in establishing the need, when using computers or manipulatives, 'to consider carefully what stands between the learners and the "knowledge" that they are intended to learn; that is, we need to focus on the learning *mediated* through employing such resources' (*ibid.,* p. 121, author's italics). Jones describes Cabri geometry (see page 60) as one of many Dynamic Geometry Environments (DGEs) which allow the learner to construct geometric objects and to specify relationships between geometric objects (Laborde and Laborde 1995) and thus provide learners with a means of expressing mathematical ideas. The concern of the study is to explore the mediation of learning in a particular teaching/learning situation involving the use of Cabri, and two factors are borne in mind throughout:

1. the inextricable relationship between psychological processes and the socio-cultural context in which they take place which means that these mental processes 'are mediated by communication that is inherently and complexedly situated' (*ibid.*, p. 122);
2. the vital importance of the move from *perceiving* to *specifying* in learning mathematics where, in the context of the study, '*specifying* requires the use of elements of conventional mathematical language' (*idem*, author's italics).

The communication of what is happening as pupils learn is being mediated by the computer environment in which, and with which, they are working, and rather than coming between them and mathematical language, it actually involves them in the use of conventional terminology.

The pupils were asked to construct the diagram shown in Figure 2.2 and to explain why the shape they ended up with was a square.

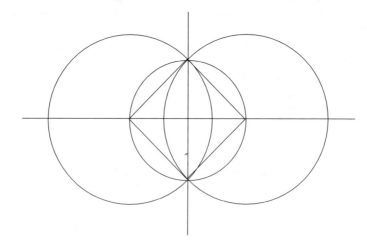

Figure 2.2: Diagram to be constructed by pupils. (Jones 1997, p. 123)

Some of the dialogue between the teacher/researcher and the pupils is recorded and an example of how they arrive at the correct mathematical terms is given when, in attempting to 'make a point in between there' they choose 'intersection' from the menu. The teacher intervenes and says 'An intersection will only give you the point where two lines cross. But there is something else which will give you something that is halfway between' (*ibid.*, p. 124). They look under 'Construction' on the menu again and both immediately identify 'Midpoint'. Jones (1997) notes that what is important is the *nature* of the interventions during the session because, as he says, 'in every attempt to reveal the mathematical thinking of learners, the balance between exploration and guidance is always problematic' (*ibid.*, p. 126). He suggests that while the two pupils had sufficient fluency with the Cabri software to complete the task and devised their own strategy for carrying it out, at the same time 'the computer environment was insufficient to allow students to fully articulate their specification in conventional mathematical language' *(ibid.,* p. 127). The final explanation of why the shape generated was a square had to be determined by them and, in the process of doing so, they were at the same time developing their ability to use mathematical language effectively in order to produce a reasoned argument.

The use of Dynamic Geometry Environments (DGE) is also reported by Hazzan and Goldenburg (1997) as part of a project to develop 'an epistemological framework for dealing, working and thinking in DGEs' *(ibid.,* p. 55). This part of their work deals with 'how DGEs enable students to approach problems through a process of successive refinement'. This idea is drawn from procedures within programming and is described in the following way:

> The model of successive refinement offers a viable alternative for dealing with complexity. One starts with a simplified version of the phenomenon under study, and refines it successfully to include more and more details, subtleties and precision. Through the entire process, the student constantly deals with the whole picture, though it may be vague or imprecise in the intermediate stages. *(ibid.,* p. 50)

One of the tasks described is where pupils had to construct a square. A 'game' was made of constructing the drawing and 'dragging' was used throughout to test whether or not it could be altered at each stage. If it could be 'messed up', then they had to start again, having identified the constraint that had not been recognized. Gradually they worked through the process and analysed the 'conditions for squareness' and added or removed features as they proceeded *(ibid.,* p. 52). It was concluded from interviews with pupils after the session that definitions of shapes in themselves were not enough for them to understand the properties that go to make up the 'whole' and using the DGE had allowed them to articulate the nature of the mistakes they had made and to identify the correct properties in doing so. The researchers are cautious, however, and note that 'the question of whether the successive refinements we see in students [*sic*] work are learning about the geometry or learning about the software, is still bothering us. The fact that the software tool does not recognize the attempt to draw a square the way any human would, and take that to be the *intent,* plays a role here' *(ibid.,* p. 55). This is an issue touched upon by Kaput (1992) in his discussion of technology and mathematics education when considering the constraints and supports the use of computers brings with it. He argues that 'whether a feature is regarded as one or the other does not depend inherently on the material itself, but on the relation between the user's *intentions* and those of the designer of the material and the contexts for its use' (p. 526, italics added). The programmer has one set of intended outcomes and an assumption of the context in which the program will be used, while the learner's intentions may be quite different and, for example, what the programmer intended to be a support may act as a constraint.

Developing spatial competences

Teaching three-dimensional shapes to 12-year-olds was the focus of an investigation by Pallascho *et al.* (1993) when they studied the effects of alternating approaches within the teaching sequence. The aim was to gain a better understanding of the respective roles of intellectual operations (analytic operations) and types of spatial representation (dynamic operations) in the development of spatialization skills in individual pupils. The study was organized around five intellectual operations identified in learning about shape:

1. *structuring* – identifying geometric properties and combinations of properties within a geometric structure;
2. *classifying* – grouping the structures together according to their properties;
3. *transposing* – establishing correlations and equivalences, bridging geometric mode with other different representational modes (physical, linguistic, algebraic, geometric);
4. *determining* – defining elements/parameters determined by geometric limitations of a spatial structure;
5. *generating* – modifying spatial structure to meet specific geometric criteria.

A case study approach was used involving two 12-year-olds (one girl and one boy) and two 14-year-old girls, all of whom achieved high scores on spatial ability tests. They spent five hours on three-dimensional geometrical activities and their work was recorded and analysed. The objective was to highlight the development of spatial competences as they worked through a series of activities in which they had to generate geometric forms and which focused alternately on analytic and synthetic competences. During analytic phases they were involved in observing shapes, and in the dynamic phases they constructed scale models. Table 2.1 is the matrix which shows the structure of the sequence of an activity.

Table 2.1 (after Pallascho et al. 1993)

Competency	Operation	Type of activity
Analytic/Synthetic	Transposing	From linguistic input to physical model
Operation	Generating	Guided construction of model
Analytic	Structuring	Recognition of characteristics of models
Analytic/Synthetic	Transposing	Physical to drawing
Analytic	Classifying	Examination of possibilities
Synthetic	Determining	Reasoning on the basis of the model and drawing
Synthetic	Generating	Unstructured problem-solving activities

The matrix indicates the alternation between the different kinds of competence the pupils had to use in the course of carrying out a task, and the nature of the operations and activities associated with them. One of the main problems found in the course of the activities was that the pupils found it difficult working with the models selected by the teacher 'which did not necessarily correspond to their perceptions of reality' (*ibid.*, p. 13). The example is given of using pieces of paper to represent infinite planes which did not appear to mean a great deal to the children. Another difficulty was that the pupils who appeared to be more familiar with two-dimensional representations of space found three-dimensional representations difficult. While Pallascho *et al.* recognize that hands-on work with three-dimensional shapes is important for generating representations of a spatial kind, 'it is also important not to create new obstacles to learning. The materials must therefore be chosen very carefully' (*idem*). An example of an activity was creating tetrahedra by sectioning a cube to obtain the fewest number of tetrahedra possible. In this example, it was noted that the 14-year-old pupils found that physically truncating the cube helped them to draw the results (which was part of the task) while the 12-year-olds did not, and 'consistently found difficulty in interpreting the drawing' required of them (*idem*).

The results from the study suggest the importance of involving pupils in carrying out activities that require the intellectual operation of *determining*. The operation of determining is when pupils are involved in 'defining the elements or parameters that are determined by the geometric limitations that apply to a spatial structure in order gradually to lead them to deductive reasoning' (*idem*). They note that 'In the case of generating a tetrahedron by means of combined symmetry operations, the students were able to discuss the role of rotation and reflection as well as the properties of the object that was created by means of those operations' (*idem*). Pupils were able to bring together what they had found as a result of their mental and tactile explorations with the three-dimensional shapes and to reflect upon these actions in order to solve a problem. It was generally considered that the study had established the importance of using the different competences implicit in Piagetian theory with respect to the development of spatial abilities in children.

IMPLICATIONS FOR TEACHING

All of the studies reported here have relevance for the teaching of spatial concepts and skills. However, an overview of the findings suggests that there are some issues which emerge as a result of considering the body of research as a whole. These appear in the four sections that follow.

Theoretical considerations – rationales for teaching

1. Bearing in mind the importance of the informal knowledge that children bring with them to school, as with number, children will have formed some concepts in relation to shape and space of which teachers need to be

aware. Some of these perceptions are bound to be idiosyncratic and either not conform to mathematical conventions or simply be wrong. If we accept that intuition plays some part in the development of spatial awareness, then this seems inevitable. It is important that children's preconceptions of space be identified in order to build on them either by extending them or to help them to achieve a correct understanding where there are misconceptions so that the foundation they build on is sound. Bearing in mind that in their earliest experiences, children recognize whole figures and do not see the parts that go to make up that whole, misconceptions may be something as simple as giving the wrong name to a shape. On the other hand, they may use the language of properties incorrectly and 'know' about corners but really be focusing on sides.

2. The research as a whole emphasizes the importance of physical activity in order for children to develop visualization skills, and this applies for younger and older children. This needs to be done in a variety of contexts as well as using a variety of materials. The use of manipulatives (pre-structured materials) can be helpful but there is a danger that children will expect all shapes to conform to these prototypes. Also, we are reminded again by the research that the mathematics is embedded in these shapes, and children need to be guided in extracting the mathematics from them.

3. The language of shape and space is the language of position as well as form. Children need practice in using this kind of language as well as the language of the properties. Experience with placing objects over and under, or turning them and flipping them, all helps to build up a repertoire that will help them interpret new experiences as they move on.

4. The notion of levels is helpful in tracking children's development of thought in connection with shape and space. What is important is not to see them as rigid and as barriers. On the other hand, referring to a sequence of the sort offered by the van Hiele levels can help to identify where a child may be coming from or where they may be going to; it can also help to understand that a child may be regressing a bit before taking the next step forward.

5. A focused awareness of the variety of strategies children may use when tackling problems of a spatial nature can provide a teaching repertoire of different perspectives children bring to these situations. The individuality of their learning will be made explicit in these strategies and is not by any means predictable, so that a conscious effort to identify them and draw them together can provide a valuable teaching resource.

6. While the active construction of concepts on the part of children is desirable for meaningful learning, it is clear that there is also a place for teacher intervention. For example, if a child is having difficulty with the idea of rotation, drawing from a repertoire of knowledge and past experience (as identified in (4) above) the child can be guided to an understanding of the concept. There is a balance to be sought between children constructing knowledge and constructive intervention on the part of the teacher.

7. The link between spatial ability and mathematical achievement raises the issue of gender differences in children's learning, particularly in relation to shape and space. Boys have been found to have better spatial visualization skills which indicates that this discrepancy has to be addressed through the learning experiences girls are given in the classroom. Girls need activities to develop their visualization skills, using both two-dimensional and three-dimensional shape not only to extend their knowledge but also to build their confidence in an area that is often interpreted as a male domain.

Teaching of spatial concepts and skills to younger children

1. Stressing again what has been said above, in the early stages in particular children should be given everyday three-dimensional objects to handle and explore, then gradually lead both to separate out the properties (e.g. sides, corners, points) from the shape and to develop the language that helps them describe, internalize and order their thinking in relation to the shape and its properties. This allows children to form concepts and the associated language that is necessary to understand shape at the more abstract two-dimensional level.
2. The levels of the development of spatialization skills provide a useful reference when considering the sequencing of children's learning and teaching of these skills. Children may be able to tell a square from a rhombus from a holistic visual image, but not be able to say why they are different. Definitions in themselves are not enough to help children understand what characterizes these differences and these are best learned by experiencing in some way *how* a square differs from a rhombus (by superimposing one on top of another, for example).

   When directing children's attention to particular properties, it appears that they do not necessarily pick out the property from the shape. This kind of activity and others that focus children's attention on the properties of shape help children to move from a holistic to a more analytic stage. It allows them both to see the difference and to experience it in a tactile way.
3. Children have difficulty with the idea of angle and the evidence that they do not associate it in a natural way with 'amount of turn' suggests that doing so early on may well cause confusion. The examples of angle in their environment which they can identify are all static in nature, and again we may think children are focusing on the angle in a situation when they may well be focusing on something else. The most obvious example of the latter is when they judge the size of angle by the length of the lines used to represent it and a measurement of the space between the lines. There is some evidence that the teaching sequence ought to begin with the static and move gradually from the static to consider dynamic examples (amount of 'opening') and static examples at the same time, eventually arriving at an idea of the amount of turn.

4. Two-dimensional shape is an abstraction and a two-dimensional array is a further abstraction, more especially when children are being asked to use such an array in a quantifying situation. Activities involving sequencing and counting will help them to organize their knowledge and develop strategies and structures for using it. For example, early attention to counting techniques and the idea of 'composites' for counting with three-dimensional shape and activities that will generally allow them to group properties systematically can be helpful.

Teaching spatial skills and concepts to older children

1. Spatial skills and concepts that were not well learned in the early years cause problems that persevere into later years of schooling. The matters touched upon in (2), (3) and (4) above are equally applicable for older children. Teaching of geometric knowledge at this level tends to rely more on definition and the research suggests that often the concepts that make up a definition (such as line or intersection) are not in themselves fully understood, so that it is unlikely that relationships defined in these terms will be understood either.
2. Children who have geometric knowledge but cannot use it in a problem-solving situation would benefit from the kind of activities mentioned in (4) above. Experience of identifying pattern, grouping properties that are related in some way and general structuring activities will help them to form schema necessary for problem-solving. Actually involving children in the development of a model for solving problems would help to direct their attention to the features that are common to all problems. Flow diagrams are one example of this kind of approach where they are involved in the process of analysis and are made to think in terms of the salient features in a problem-solving situation. This leads them to think in terms of larger units and how they are related to each other.
3. The research evidence related to teaching about perpendicularity is a good example of how easy it is to take for granted that children 'see' what we see. The three steps of identification, selection and inference are useful guidelines to use when concepts of this kind are introduced. Words in themselves describing this sort of spatial concept are, by the nature of the exercise, inaccurate because the abstraction is compounded; the perceived meanings of the words used and what is intended to be perceived in a geometric figure are both open to different interpretations on the part of pupils and teacher.
4. Identifying pupils' abilities in relation to the different kinds of competence and intellectual operations involved in a single geometric problem-solving exercise could be a useful diagnostic exercise. The kind of information gained about the extent to which they can engage in the different intellectual operations involved could provide valuable information about their strengths and weaknesses.

Computer technology

1. The generally positive outcomes of studies using computer environments in geometric teaching suggest that the technology provides a rich tool for teaching in this area of mathematics in particular. The dynamic aspect of programs is valuable both in understanding transformations and in using the drag option in the course of understanding the properties of shapes.

2. As with any other teaching situation, there is strong evidence that appropriate and well-timed teacher intervention is very important when using technology, particularly with respect to understanding the mathematical terminology used.

3. Teacher awareness is particularly important in using a Logo type of environment to develop the concept of angle with younger children where misconceptions, once established, have been found to be particularly difficult to dislodge.

4. It seems possible that computers may be well suited in particular to develop pupils' ability with respect to deductive reasoning and the intellectual operations involved in defining shape.

5. Gender issues may well need special attention in a computer learning environment, especially with older pupils when girls' attitudes may be changing with respect to involvement in what may be seen as a male domain.

Chapter 3

Probability and Statistics

INTRODUCTION

In 1982 the Cockcroft Committee reported that 'Surprisingly few submissions which we have received have made direct reference to the teaching of statistics' (Committee of Inquiry into teaching Mathematics in Schools 1982, p. 234). They also note that 'the increasing use which is being made of statistical techniques in so many fields makes it highly desirable that those who study mathematics [at a higher level] should have some understanding of probability and statistics' (*ibid.*, p. 172). The difficulty of achieving this is reflected in a study by Glencross (1998) reporting the development of a 'statistics anxiety rating scale' which is used with students who are described as 'apprehensive' when faced with statistics sources in the social sciences and generally approach these with a negative attitude (*ibid.*, p. 256). The importance of the curriculum providing the activities that allow children to develop these mathematical skills was acknowledged and since that time at least some probability and statistics has been included at both primary and secondary levels in the UK. This was not necessarily the case in other countries such as the US (see Shaugnessy 1992) but it is a subject which has gained momentum over the years as being one which provides a real-life context in which arithmetical and algebraic skills, as well as spatial and representational skills, can be developed. It has had different guises at primary level (much of it embedded in pictorial representation) and currently appears in the UK National Curriculum as Data Handling.

The introduction of the subject means that teachers themselves have to acquire new content knowledge as well as new teaching skills to meet the demands of the curriculum. However, the body of research related to it is sparse and the relative newness of probability and statistics (known together as stochastics) in the curriculum may account for this sparseness to a large extent. It may be that until it is given a firmer focus generally within mathematics curricula, it is unlikely to attract a substantial body of study.

A brief survey of the research indicates that there has been some work carried out with respect to probability and its teaching, to pupils' concepts of probability, what they bring to learning situations that involve the element of chance, and the notion of predicting outcomes. However, there appears to have been little done in relation to specifically statistical concepts such as the notion of average or the relative accessibility of, for example, the different kinds of average to children. The studies reported in this chapter reflect this imbalance in research on stochastics in school curricula and also the paucity of relevant studies. For this reason, the chapter will take a slightly different format. We shall first consider briefly recommendations as to what should be taught in the stochastics curriculum, then examine the nature of stochastics, theoretical considerations related to probability and some of the research that has illuminated these aspects of the subject. A consideration of a sample of studies carried out in schools will follow and, finally, implications for teaching.

WHAT PROBABILITY AND STATISTICS SHOULD BE IN THE CURRICULUM

In 1984 one of the topic areas under consideration at the Fifth International Conference of the Group for the Psychology of Mathematics Education (ICME 5) was the teaching of statistics, and the following statement was made:

the following trends concerning the teaching of probability and statistics at the school level may be identified:

- more emphasis on statistics than on probability with special emphasis on descriptive statistics including the use of exploratory data analysis methods;
- emphasis on applications and model building;
- use of simulation both as a practical and didactic tool:
- use of calculators and computers;
- use of project work. (Rade 1984, p. 300)

One participant noted that 'students can learn to graph, summarize and interpret data without first studying probability' (*idem*). Referring to geography, economics and the biological sciences (including psychology), another said 'Statistics has been included in these courses not as a result of pressure from statisticians but because the subjects themselves are becoming more numerical. This is the desire to quantify concepts and measure effects which were previously only discussed qualitatively. It is this desire to use real data that motivates the students and makes the study of statistics important to the subject area' (*idem*). This was followed by an appeal for relevant courses for 16–19-year-olds which:

1. should be oriented more towards data than probability;
2. include a background in reading data and tables;
3. ensure that the statistics arise in real contexts. (*ibid.*, p. 303)

Eight years later, Schaeffer (1992) states that 'Data ("numbers with a context") provides a concrete basis for development of mathematical ideas and serves as a way to connect mathematics to the lives of the students' (p. 286). He goes on to suggest that probability should *not* be the basis for the teaching of statistics at school level, but that the notion of randomness should be established by giving children experience of 'long runs' using manipulatives. Rather than an 'add on' approach of new materials to the mathematics curriculum, the statistical aspect of the content should arise from questions asked by children based on data they themselves have collected. Schaeffer cites as an example the approach used in Japanese industry: 'This educational process emphasizes statistical sense rather than techniques, application ability rather than systematic knowledge, and problem-oriented rather than technique-oriented learning' (*ibid.,* p. 287).

These views will be an interesting background against which to consider what follows with respect to school curricula and what the research reflects.

PROBABILITY AND STATISTICS AS A TEACHING SUBJECT

Shaugnessy (1992) identifies the problem that teachers have in the teaching of stochastics which arises from the subjective viewpoint that they, as individuals, bring to the subject and the perceptions and expectations that determine their attitudes towards it. This arises from the fact that the subject is one which is based on gathering data and the *interpretation* of that data. On the one hand, an hypothesis may be formulated before the data are gathered and determine what factors are to be observed and recorded, or alternatively, data may be gathered randomly, and represented and interpreted, and the perceived pattern then forms the basis of an hypothesis. Whatever the situation, there will be an element of interpretation both within the process and at the end of it, and interpretation necessarily involves an element of subjectivity. As Shaugnessy (1992) puts it, 'Intuitions, preconception, misconceptions, misunderstandings, non-normative explanations – whatever one might call them – abound in the research on learning probability and statistics' (p. 465). Of all the topics taught within mathematics, stochastics is the one where teachers are becoming more aware of the perceptions and opinions their pupils bring to it (perhaps from a recognition of their own subjectivity in this respect). Teachers' own frameworks may not be compatible with those of their pupils which, as Konold (1991) suggests, makes communication about the subject problematic. It is perhaps the least 'pure' mathematical topic in that sense, and hence can be the most difficult to teach for teachers who have become imbued with the idea of mathematics as objective and characterized by definitive correctness. Shaugnessy (1992) gives an historical overview of the development of this branch of mathematics and brings in arguments about the duality of 'science' versus 'belief'. Stochastics is identified as a 'low science' because it is not founded on absolute truth but on empirical evidence. It is perhaps this aspect of stochastics that has made it reasonably accessible to other teachers who feel more secure with this view and these areas of mathematics that have their roots more obviously embedded in the 'real' world and human activity around them.

Statistics deals with everyday situations such as the favourite colour of cars, the lifespan of people living in a specific area within a country, how large plants of a different type grow under given conditions, or the size of an egg. Each of these situations requires the gathering of data, and whatever the focus of that data, the outcome will be in the nature of a *tendency* of some sort rather than a statement of *fact*. Statements such as 'People are more likely to buy a white car than any other colour' are the kinds of outcomes that arise from gathering and interpreting the collected data. It is this predictive characteristic of statistics arising from this kind of data and the extrapolation of information from it of a probabilistic nature that brings the two together, although probability also has a somewhat purer mathematical pedigree, as we shall see.

PROBABILITY

A brief résumé of the three types of probability is given by Koirala (1998). Although they may not all necessarily be of relevance to school curricula, they provide an insight into the different interpretations given to probability and help us to understand how the variety of misconceptions that exist arise in the first place. A description and definition of each type is offered.

Classical probability is based on assuming equally likely possible outcomes when carrying out an event (e.g. flipping a coin). The original definition of Laplace is given as 'the ratio of favourable cases to the total number of equally possible cases' (*ibid.*, p. 135). The word 'possible' however, is open to many interpretations and a refined definition (proposed by Kline 1967) is 'if, of n equally likely outcomes, m are favourable to the happening of a certain event, the probability of the event happening is m/n' (*idem*). Shaugnessy (1992) refers to classical probability as assigning a probability in an experiment when all outcomes are equally likely.

A *frequentist* perspective of probability is one based on the observed frequency of an event occurring out of a given number of trials (e.g. how many 3s will occur in 500 rolls of a die?). The definition given is 'the ratio of the observed frequency to the total number of trials in a random experiment over the long run' (Koirala 1998, p. 135).

A *subjectivist* perspective is described by Borovcnik *et al.* (1991) as 'evaluations of situations which are inherent in the subject's mind' (p. 41). This is the kind of probability we engage in from day to day when we estimate the likelihood of an event happening from our personal observations, but which is open to alteration as a result of our ongoing observations and re-interpretation of these events. (Because of the dependence on personal experience, it is sometimes referred to as *personal probability*.) Shaugnessy talks of 'degree of belief' in connection with this viewpoint, again emphasizing its relativity and dependence on the individual in its interpretation. (*ibid.*, p. 469). Konold (1991) elaborates further on the complexity of these views. He refers to such personal theories as being normative theories which specify 'how "rational" people ought to formulate and alter their beliefs in the light of new information' and suggests that they are not descriptive theories of 'how people actually formulate and alter subjective probabilities' (p. 143). He is highlighting the assumptions made in attempting to define probability in this way and there is the implication that people do not always behave 'rationally'.

Shaugnessy (1992) suggests that much research in the past has been taken up with establishing the superiority of one perspective of probability over another. However, where mathematics education and the teaching and learning of probability and statistics are concerned, he states that 'a modeling point of view is a much healthier perspective from which to discuss the different views of probability' (*ibid.*, p. 469). He comes to this conclusion in the following way:

> There are indeed, certain normative ideas of stochastics that we want our students to understand and to be able to use. However, while some probability experiments can best be modeled by a uniform probability space, others can best be modeled from a frequentist perspective. There are problems in which a 'marriage' between experimental frequencies and classical theory is desirable (such as tossing dice or using equal area spinners). There are other problems in which either a theoretical solution does not yet exist, or is not readily available to our students. In such cases, a frequentist approach has great merit. There are also probability problems in which the conflict between subjectivist tendencies and classical theory can be resolved by taking a frequentist point of view and running a simulation. Thus, I would advocate a pragmatic approach which involves modeling several conceptions of probability. (*idem*)

Psychological considerations

The fact that subjectivism is readily acknowledged within the study of probability and statistics also suggests a need for greater awareness of the psychological factors that come into play in teaching and learning this aspect of mathematics. This helps to pinpoint where subjectivism creeps in and our personal viewpoint becomes biased. Although many people may claim to be statistically naive, they would at the same time be quite open to making judgements about what is likely or not likely to happen. They do this on the basis of an heuristic (i.e. how they teach themselves or learn to make such judgements) which is developed as a result of personal experience and observation. Konold cites the work of Kahneman and Tversky (1972) who tried to explain why people's judgements tend to differ from accepted probability theory. Their suggestion was that 'because of limited information-processing capabilities, people use various judgment heuristics that allow them to summarize large amounts of data and quickly arrive at decisions' (Konold 1991, p. 145). Much of past research literature has been concerned with the two heuristics that dominate this aspect of learned mathematical behaviour, and these are *representativeness* and *availability*.

Representativeness
One of the dominant factors individuals are likely to consider in estimating the probability of something happening is the extent to which they think it is representative of the population in which it is occurring. Shaugnessy (1992) gives an example of this aspect of representativeness where the population consists of the children in a family and the sequencing of Girls and Boys within a family of six children. He suggests that many people will believe that the sequence BGGBGB is more likely to occur than either BBBBGB or BBBGGG. The sequence BGGBGB is seen to be more likely because it reflects closely the 50–50 possibility of a boy or girl being born, which the sequence BBBBGB does not. Similarly, the sequence BGGBGB appears to reflect the randomness of the process of having children, better than the sequence BBBBGB does. From a theoretical point of view, each of the 64 possibilities is equally likely to occur. What people tend to believe is that 'a sample should either reflect the distribution of the parent population, or that a sample – perhaps even a single outcome – should mirror the process by which random events are generated' (*ibid.*, p. 470).

A different sort of reliance on representativeness is shown when individuals tend to ignore the size of the sample from which an event is drawn. They may think it equally likely to get four heads from tossing a coin ten times as it would be to get 40 heads from tossing a coin 400 times. They are unaware of the fact that extreme cases, such as a run of four heads, is more likely to occur in a smaller sample of events than a larger sample. Shaugnessy gives other examples of how representativeness is manifested in people's judgements about the probability of events occurring. One is the 'negative recency effect' used by gamblers when they believe that after a run of heads, a tail is bound to occur next. Another is the 'base-rate fallacy' where Shaugnessy gives the example of people predicting the likely profession of a person, having been given a couple of 'base rate statements' about them. Having been given such statements describing an individual from a sample of 30% engineers and 70% lawyers, they will guess that the person is an engineer and not a lawyer because the 'base rate statements' fit with their personal stereotype of an engineer. Their judgement is based on the descriptive statements which they assume from their personal experience to be representative of the attributes of a population (in this case, engineers) and they ignore the fact that only 30% of the sample is composed of engineers. This is an example of how subjectivity enters into probability and is influenced by personal experience, and qualitative aspects of judgement-making dominate the quantitative.

Availability
Availability is a simpler heuristic affecting an individual's judgements about the probability of something happening and refers quite simply to the experience an individual has had as opposed merely to observations they may have made. As Shaugnessy states, 'We all have egocentric impressions of the frequency of events' and these will dominate our judgements about the likelihood of an event occurring (*ibid.,* p. 472). The example is given by Truran (1998) of children who, when throwing a die, are convinced that they are less likely to throw a 6 than any other number. Truran puts this down to the fact that in many board games, children are required to throw a 6 before having a 'go' and they recall long waits in doing so. If it takes a long time to throw a 6, there must be less chance of throwing it than any other number. This is the 'available' relevant experience upon which they draw in judging the probability of throwing a 6.

Another example of how availability as an heuristic comes into play is given by Shaugnessy when he discusses more complex examples involving combinatorics. An individual's success in this kind of task will depend upon an individual's level of ability with number. Given a group of ten people and asked whether it is possible to make up more pairs of people than groups of eight people, more of those questioned will say it is possible to form more pairs as opposed to groups of eight, simply because they find it easier to work with the number 2 than 8. It is less complicated and easier to put two people together than it is to gather together different arrangements of eight.

Other misconceptions

Conditionality
Apart from the misconceptions and error that can arise from the heuristics we use as individuals in our learning about probability, there are other misconceptions that have been identified by research. A major one of these is *conditionality* and is of some concern in the teaching and learning of probability and statistics.

Shaugnessy cites the work of Falk (1993) who identifies four reasons why conditionality in probability problems causes particular difficulty for pupils.

1. Pupils may have difficulty in deciding what the conditioning event is within a problem. The following example is given:

    > There are three cards in a bag. One card has both sides green, one card has both sides blue, and the third card has a green side and a blue side. You pull a card out and see that one side is blue. What is the probability that the other side is also blue? (*ibid.,* p. 474)

    Results of using this problem have shown that pupils focus on the conditioning event as being a *card* whereas it is really the *sides of the card* that should be the focus.

2. Pupils may confuse conditionality and causality. There is a difference between the probability of having measles when a rash appears and the probability of having a rash because one has measles. Having a rash does not depend upon having measles. On the other hand, measles does cause having a rash.

3. Conditionality may become confusing when the conditioning event takes place *after* the event it conditions. The following experiment is given as an example of this kind of situation.

> An urn has two white balls and two black balls in it. Two balls are drawn without replacing the first ball. (1) What it the probability that the second ball is white, given that the first ball was white? (2) What is the probability that the first ball was white given that the second ball is white? (*ibid.,* p. 473)

As Shaugnessy says, it is easy to see why the probability for the first event is 1/3 , but not so easy to accept that the probability of the second question is also 1/3. In this case, the first event is contingent upon the second which seems to contradict rationality and is difficult for pupils to accept. (It is suggested that physically acting this out will help pupils to understand how the second event *can* be the conditioning event.)

4. The language of conditionality confuses pupils, and what Shaugnessy calls the 'cognitive overload' it brings with it can lead to misconceptions. The example of the urn and balls above is indicative of the kind of problem pupils are often faced with, and although it is succinctly worded without the use of conditionals such as 'if', there is a sequence of events that has to be remembered in order to reach a solution.

Piaget and probability

Piaget and Inhelder (1975) saw the development of the concept of probability by children as taking place in accordance with his developmental theory in stages that have been supported by later research (Shaugnessy 1992). Up to the age of 7, children are not able to distinguish between what is a necessary and what is a possible event, and they have no concept of uncertainty. During the next years and up to 14 years of age, while the child may be able to distinguish between necessary and possible events, they are unable systematically to produce a list of all possible outcomes. The theory holds that they are also unable to make an abstract model of a situation of this type since they are supposed not to have developed all the necessary mathematical skills. After 14 years of age and when their number skills have developed accordingly, the child begins to understand probability in terms of the 'limit of relative frequency' and since this is identified in terms of a ratio, the concept of fractions plays an important part in the development of the concept of probability.

Green (1983) carried out a longitudinal study with more than 3,000 children aged 11 to 16 in the UK to see whether their development of the concepts associated with probability matched the stages described by Piaget. In his study, the tasks pupils were engaged in included work with tree diagrams, spinners, various visual representations of randomness, and marbles. Tree diagrams were not easily understood nor the multiplication of fractions in calculating probabilities. He concluded that:

1.   the concept of ratio is vital to children's understanding of probability;
2.   the level of understanding of the language of probability is poor (e.g. words such as 'certain' and 'least');
3.   a systematic approach to the teaching of probability and statistics in schools is necessary to overcome children's misconceptions in connection with the subject.

Konold (1991) builds on Fischbein and Gazit's (1984) work in connection with intuition and probability and the belief that intervention (i.e. teaching) can improve pupils' intuitive ideas about probability. He argues that for the teaching of probability to be effective, teachers will not only have to be aware of children's intuitions in relation to probability, but also to devise means of getting them to examine these intuitions from three different perspectives:

the fit between their beliefs and
1.   the beliefs of others;
2.   their other, related beliefs;
3.   their own observations. (Konold 1991, p. 152)

One approach with respect to the first of these is to engage children in class discussion which allows the presentation and challenging of different viewpoints. In the second case where children are encouraged to examine the consistency of their own beliefs, it is suggested they should be motivated to ask questions in relation to these such as 'Why do you think . . . ?', 'How can you explain . . . ?' variety. In the case of the third point concerning children's own empirical observations, Konold argues that there are two things which people generally do not do which weaken their empirical observations to a considerable extent:

1.   they do not keep accurate records;
2.   they do not look for data that would be inconsistent with, and thus controvert, a belief they hold. (*ibid.*, p. 154)

He suggests that to overcome misconceptions about probability, these two features should be emphasized in the classroom teaching of the subject. The notion that intervention can be effective is supported by Garafalo and Lester (1985) who consider that metacognition includes knowledge of self and that this can be regulated just as with any other kind of knowledge. Hence it is possible to correct one's own misconceptions with respect to probability as is the case with other conceptual areas.

Grade 3 pupils needing help with having the frequency explained on the vertical axis. Three had difficulty with Question (3) and the other six made up stories to justify their answers.

In the *fairness of dice* activity pupils were asked if one number was more likely to come up than another when dice are thrown and then had the notion of *fair* and *unfair* dice explained to them. The pupils in Grades 3 and 6 were given two dice to consider and Grade 9 pupils had three. One die was fair, one loaded and the third (Grade 9) had two of each of the numbers 1, 2 and 3 on it. The researchers note that the dependence on perception is one of the characteristics of the mode of iconic functioning which they expected to play a part in this experiment. Over half of all pupils believed that some numbers came up more often than others, but a couple in Grade 9 were not certain. Ten of the pupils used the physical characteristics of the dice to make their judgement. Only six pupils were able systematically to carry out trials to determine the fairness of the dice, two in Grade 6 and four in Grade 9. The interpretation was that those making judgements using physical characteristics were 'mainly relying on iconic features rather than logical concrete symbolic reasoning' but iconic features were not so much in use in the higher level responses (*ibid.*, p. 374). This differed from the first two activities where 'iconic support of different types occurred along with the highest level responses' (*idem*).

The results of this study are interpreted as supporting the problem-solving path identified by Collis and Romberg (1991) which traces decision-making down one of two paths, iconic or concrete symbolic reasoning. They note that some pupils took entirely one or other of the two paths while some interwove paths using both, and that all possible paths were covered by the variety of pupil responses. The second important finding they identify from the study is 'the appearance of different specific strategies according to the context of the protocol set' (*ibid.*, p. 376). They conclude that:

> For mathematics educators, particularly those embarking on studies of the teaching and learning of chance and data, we must know *both* the underlying structure *and* the specific courses of action which are likely to be chosen, if we are to make intelligent suggestions for the curriculum and methods of instruction. (*idem*, author's italics)

The results of this investigation have made a contribution in this respect by showing how pupils approach three situations fairly typical of those many teachers use.

Random generators

Truran and Ritson (1997) have investigated young children's understanding of Random Generators (RGs). The study was concerned with a comparison of the role RGs play in children coming to understand probable outcomes and investigated the possibility suggested by Bramald (1994) that children tend to assume an underlying symmetry of outcomes because of the RGs used in the teaching of probability in the earliest stages (e.g. dice and equal-area spinners). They approach probability expecting all outcomes of an event to be equally possible. Truran and Ritson (1997) state that 'The physical properties of RGs, and the physical arrangement of RGs both have significant influence on children's perceptions' (pp. 239–40). The studies reported were carried out in a variety of schools in Australia and in Northern Ireland. The latter study involved children of 5 to 9 years of age and their perceptions of the likely outcomes using a variety of RGs were recorded. The Australian study involved using group tests in a classroom situation with 300 children from Years 5 and 7, and again the focus was the behaviour of the children when using a variety of RGs.

Both studies involved extensive interviews with randomly selected children from the samples. One of their findings was that the physical properties of RGs affected children's responses to questions asked. For example, when asked whether there was a greater chance of obtaining a 6 from rolling a die or from selecting the disc with '6' on it from a collection of six discs numbered 1 to 6, one answer was 'The discs are better because they are separate and can all jump about in the box, but the numbers on the dice can't move' (Truran and Ritson 1997, p. 242). Another example involved children being shown an urn with six numbered tennis balls and a six-sided die, and most children predicted that the likelihood of choosing a particular number would be different in each case. Tennis balls were thought to be liable to 'roll into a corner' and hence be less accessible (*ibid.*). The general conclusion reflects once again the point made by Shaugnessy that children bring preconceptions, which are often misconceptions, to the study of stochastics that must be unmasked in order for them to learn effectively.

Graphing representation

Ainley *et al.* (1997) report a study which formed part of the Primary Laptop Project. By giving pupils of this age (5- to 11-year-olds) liberal access to laptop computers, the aim of the project is to involve teachers and their pupils in their use in order to:

(a) study the effect on children's mathematical development;
(b) study the general effects of use of laptops in this way on the school, teachers and children; and
(c) to provide student teachers with an insight into the educational benefits of using laptop computers.

One laptop was allocated to two or three children. Early work within the project focused on using the laptop to teach children graphic skills within the context of work on spreadsheets. This choice was based on some evidence

(Ainley 1995) that children's difficulties in this area may largely be due to pedagogic and not cognitive reasons.

The process pupils used in this study was 'active graphing' and the purpose was to investigate features 'which are particularly effective in supporting children's ability to interpret graphs showing relationships between two variables' (Ainley *et al.* 1997, p. 2). Preliminary studies (Ainley and Pratt 1995b, Pratt 1995) indicated that there are four main factors to be attended to for a graphing activity to be effective.

1. It should *be experimentally based* involving practical work which generates data.
2. The purpose of the activity should be *meaningful* to the pupil so that (a) they have 'ownership' of it, and (b) it offers useful information that otherwise would not be available.
3. It is important that the pupil has *control over the independent variable* and is able to change its value at particular stages of an experiment (e.g. the length of side of a rectangle when exploring its relationship to area of the rectangle).
4. The activity should be *repeatable many times* to allow for the exploration of the data in both numerical and graphical form.

The researchers note psychological characteristics of importance in involving pupils in work of this kind. The first of these was the recognition that the pupils would bring *intuition* and informal ways of looking at things that would 'shape their interpretation of their experiments and their data' (*ibid.*, p. 4). They also considered *phenomenological impact*. This is exemplified in the study on area as being low because such an experiment was not in itself considered likely to have much impact on the pupils' conceptions of the size of area as this is a difficult judgement to make based on visual inspection alone. The third characteristic they viewed as important was the notion of *metaphoric resonance* which was the possible connections a pupil could make between the physical experiment and the graphical images they produced related to it.

Two groups, each with two pairs of children, were chosen for close observation and taped interviews, on the basis of whether they were judged to be open in discussion rather than on their mathematical ability. One of four activities designed is reported here and involved the pupils in designing a display board using a border 75 cm long and having as large an area as possible. The children used a border of this length, and in the preliminary stage of the study they used it to explore different possible rectangles they could make. Significant incidents were selected for analysis from this data and children's strategies were analysed and categorized.

Having carried out their experiment physically, the children began work on the computers by drawing rectangles. They then entered the lengths, widths and areas of a number of rectangles in three columns of a spreadsheet on the computer. The kinds of characteristics of the graphing activity that were identified and coded throughout this procedure include trend-spotting, shape-spotting, identifying maximum values, changing preconceptions, spotting inaccuracies and testing hypotheses. This is an ongoing project but the results reported give a clear indication of the powerful effect this computer environment has on the thinking of the children. They have, for example, immediate visual access to an example of a 'maximum' in graphical form which they learn can be changed by altering a single variable. They also have the tabulated numbers alongside the graph which present a pattern and allow them to anticipate and predict what will happen by altering a dimension. Numerically, they are coming to understand that to achieve the maximum for which they are searching, they have to consider a length which is not a whole number. They gradually realized that the answer was not related to a whole number but to a 'half or a quarter' and eventually arrived at a length of 18.75 cm. The pupils appeared to be gaining all the benefits of using computers in this context that Dunkels (1992) identified as important mathematical concepts to be learned within the study of stochastics generally.

Fairness

Pratt and Noss (1998) have also used computers to explore primary children's statistical thinking in which their focus was probability and the notion of 'fairness'. In their study, they viewed the computer as a research tool as opposed to a teaching tool. They viewed it as enabling the researchers to focus on children's construction of meaning for stochastic phenomena, in this case exploring children's intuitions about fairness and randomness. They note that 'our appreciations of stochastic phenomena live in two almost distinct worlds: the one "everyday" and the other "mathematical"' (*ibid.*, p. 17) and they suggest that it is in the intersection of the two worlds and the thinking associated with them that mathematical meaning-making can be found.

   In their study, they report work carried out with 10- and 11-year-olds who used a 'dice' gadget which was devised to allow the learner to 'throw a die' any number of times. 'Gadgets' in this context are seen to embody 'quite explicitly and accessibly a mathematical representation, of how it works' (*ibid.*, p. 18). The representations are seen to be *instructive* insofar as pupils have to make sense of them and *constructive* in the sense that pupils can use them to extend their knowledge. The actions of the two girls described in the study were monitored and their dialogue recorded and analysed as they worked with the computer trying to find out how the 'dice' worked. Several 'slices' of their talk are recorded which indicate the development of their understanding of the gadget.

   One outcome of this part of the research was the identification by the two girls of three aspects of their intuition about fairness that arose in their thinking:

1. if the dice is fair, you can't tell what comes next;
2. if the dice is fair, there are no obvious patterns in the sequence of results;
3. if the dice is fair, the different outcomes should be equally, or nearly equally distributed. (*ibid.*, p. 22)

Pratt and Noss use two notions to interpret what was happening as the girls interacted with the microcomputer. One is *phenomenological primitives* (p-prims) (di Sessa 1983), which refer to their intuitions related to sense-making as they proceeded through their investigating; the second is the notion of *webbing* which is used to describe how the internal and external resources support the learners in their learning in the microcomputer environment. They noted the following resources as crucial in the development of the girls' understanding:

- the repeat primitive, which supported a move towards aggregation of results;
- the graphing tool *which supported the visualization of the results in terms of qualitative variations (e.g. there were too many sixes);*
- choose-item, the primitive we designed to articulate explicitly the notions of random choice among a set of numbers;
- the *workings box*, which provides a living representation of the distribution (or, at least, a way of thinking about the distribution). (*ibid.*, p. 24, authors' emphasis)

They conclude that the thinking of the two pupils had evolved to a point where data and distributional meanings for fairness and randomness had become incorporated in it, and they would be able to call upon these new meanings in future learning situations concerned with stochastics.

STATISTICS IN THE SECONDARY SCHOOL

Probabilistic thinking

An investigation by Amir and Williams (1994) reports the effects of culture on children's probabilistic thinking by age 11. Their sample consisted of thirty-eight 11- to 12-year-olds in their first year of secondary school who were interviewed to explore their beliefs and preconceptions about concepts such as 'chance' and 'luck'. The culture of the children is defined as 'their language, their experience and their beliefs' (p. 26). The researchers describe their theoretical standpoint as:

to view attitudes, belief, experiences and use of language as the means by which culture might influence children's probabilistic thinking (in particular their attribution of events to chance, use of heuristics, level of understanding, and use of an outcome approach). (*ibid.*, 1994, p. 26)

They draw on the work of Fischbein *et al.* (1991) and Saxe (1991) amongst others, who have also studied cultural influences on probabilistic thinking. They also refer to the availability factor (Shaugnessy 1992) and describe it as being circumscribed by the strength, frequency or recency of relevant experiences within the culturally determined aspects of pupils' lives (e.g. gambling and the use of board games).

The interviews with pupils involved discussion about language associated with probability, beliefs, superstitions and attribution, background and experience related to chance, and specific questions related to probability. These were analysed and the results classified under the headings of 'language', 'beliefs', 'experience' and 'probabilistic thinking'.

The results of the study showed that pupils' beliefs affected their views on probabilistic thinking more than any other factor, and these were beliefs not only associated with religion but with superstitions and attribution generally. With respect to accidents, for example, some saw these as acts of God, others saw chance playing a role, but most saw them as controllable by people's behaviour. When it came to manipulatives used in studying probability such as dice, 'quite a number of children thought in different degrees of certainty, that their results depend on *how you throw*, or handle, these different devices' (Amir and Williams 1994, p. 28, authors' emphasis). There were no marked differences in pupils' experience of games involving probability. They were found to apply the availability heuristic in the way in which they viewed dice and considered it 'harder' to throw a 6. They also employed representativeness and believed that, in tossing two coins, you would be more likely to get one head and one tail rather than two heads or two tails. A majority employed the outcome approach when it came to forecasting weather when '16 out of 21 interviewees answered that if after a prediction of 70% chance of rain it did not, in fact, rain – then the probabilistic prediction was wrong' (*ibid.*, p. 29). Explanations of some events appeared to indicate an 'equiprobability bias' where, for example, in a sample of thirteen boys and sixteen girls, the chances of either a girl or a boy winning were seen to be equal.

While acknowledging the influence of ability on the probabilistic thinking of an individual (Green 1982), the conclusion is drawn that cultural factors do affect this kind of thinking in pupils in a way that leads to different types of reasoning. These are identified as superstitiousness, religion, causality and determinism, tricks and suspicion, and equiprobability. This analysis is helpful in pinpointing more specifically some of the factors identified by Shaugnessy (1992) that contribute to the subjectivism individuals bring to the study of probability and statistics.

Models

A study carried out by Doerr (1998) explores pupils' thinking about multiplicative structures underlying exponential functions (Confrey and Smith 1994) and the role that probabilistic events play in relation to these functions. The theoretical framework draws on the use of models and is described in the following way:

> A model building approach to learning mathematics suggests that an important goal for learning is for students to be able to construct mathematically significant systems that can be used to describe, explain, manipulate, and predict a wide range of experiences. (*ibid.*, p. 256)

The stages of constructing a model are described as:

1. the *elicitation* stage when the pupils are faced with a situation where a model needs to be constructed;
2. the *exploration* stage where the model constructed is explored and studied for its general applicability, in order to extend its power, and to make the underlying structures explicit;
3. the *application* stage where new situations are found in which the model may be used.

An activity was designed to be used with two classes of seventeen and thirteen pupils in Grades 10 to 12. The activity was one through which pupils would be led to construct their own model arising from a problem situation. The classes were videotaped and extensive field notes were taken. The problem was stated as follows:

> You will start with one or more M&Ms in your cup, shake the cup and pour the contents onto a napkin. For each M&M that has the 'M' showing, add a new M&M to the cup. Put all the M&Ms back in the cup and repeat the procedure 10 more times. For each trial, record the total number of M&Ms in the cup. (*ibid.*, p. 259)

The pupils had to graph the data and find an equation that they could use to predict the total number of M&Ms for any single trial, and this was to provide an introduction to a *growth model*. A second problem situation used the same materials but involved beginning with a full cup and removing M&Ms. This was their first introduction to 'exponential decay' and was an attempt to ground their thinking 'in an everyday experience that was familiar and readily understandable' (*idem*) and to understand a *decay model*. The third activity was abstract in the sense that pupils were asked to imagine a cupful of four-sided M&Ms with each having only one 'M' and to carry out the previous two activities using this.

They moved through stages where they focused on counting and recording and moved on to conjecturing and reasoning about why some equations did not work, and eventually arrived at a point where they could use their data table and their equation to 'unproblematically make several predictions about the behavior of the system' *(ibid.,* p. 262). The results indicated that both groups of pupils had 'created meaningful interpretations of the probabilistic growth situation' and had been able to abstract and generalize their findings to a hypothetical experiment for the growth situation *(idem).*

University students' statistical thinking

Batanero, Godino and Estepa (1998) have studied university students' understanding of statistical association over a period of seven years, mainly focusing on the concept of association. While we are concerned here with school mathematics, the approach they adopt in their work is one that equally well could be used in schools and provides some insight into how relevant concepts are developed. Their work builds on earlier studies carried out with 18-year-olds and reported by Batanero *et al.* (1996).

They state that 'The main goal in studying association is to find causal explanations, which help us to understand our environment' (Batanero *et al.* 1998, p. 22). They recognize, however, that association does not necessarily in itself imply cause, and acknowledge the possibility of being misled by the concurrence of factors which appears to produce a correlation between some or all of them. They differ somewhat from Fischbein (1987) by adopting the view that an ability to judge association effectively is not one that is developed intuitively; rather, they recognize the fact that individuals will base their judgements on their beliefs about relationships that 'ought' to exist rather than on an interpretation of evidence they may have before them. The theoretical framework used is one that is dependent upon knowing and understanding mathematical objects. These mathematical objects are seen to emerge as a result of an individual's mathematical problem-solving activity and are linked to particular problem fields.

Students were given a questionnaire with ten items including $2 \times 2, 2 \times 3$ and 3×3 contingency tables, scattergraphs and a comparison of a numerical variable in two samples. The researchers state that sign and strength of the association and the relationship between the problem context and students' prior beliefs were taken into account when designing the questionnaire. Each item was analysed for the types of association perceived by students, direct association, indirect association or independence, and their solution strategies were classified. Following this, their misconceptions were categorized.

An example of a question is given below:

In a medical centre 250 people have been observed to determine whether the habit of smoking has some relationship with a bronchial disease. The following results were obtained.

	Bronchial disease	No bronchial disease	Total
Smoke	90	60	150
Don't smoke	60	40	100
Total	150	100	250

Using the information contained in this table, would you think that, for this sample of people, bronchial disease depends on smoking? Explain your answer. (Batanero *et al.* 1998, p. 224)

Misconceptions in interpreting this kind of evidence were classified into four types.

1. The first is a *determinist* conception of association which is based on a belief that there are no exceptions to the existence of a relationship between variables, and students 'expect a correspondence that assigns only a single value in the dependent variable for each value of the independent variable' (*ibid.*, p. 225). An example of this type of misconception might be the suggestion that there is no relationship because some people who have bronchial disease are non-smokers, while others do smoke.

2. The second type of misconception they called a *unidirectional* conception of association. This occurs when the student perceives dependence only when it is positive (i.e. direct association) and as a result, when a negative sign appears (an inverse association), they treat this as independence. One statement illustrating this was as follows: 'I personally think there is no dependence, because if you look at the table there is a higher proportion of people with bronchial disease in smokers' *(idem).*

3. The third type of misconception is called the *local* conception of association where only part of the data is used to make a judgement, as in the following statement: 'There is dependence on smoking in having bronchial disease, because if we observe the table, there are more smokers with bronchial disease than non-smokers: 90 > 60' *(idem).* In some instances of local misinterpretation, students would focus on a single variable (e.g. only considering people with bronchial disease in the example given above) or even a single cell in the table, often the one with the greatest frequency. This supports the findings of Konold *et al.* (1997) who found evidence of similar misconceptions and ascribed them to an inability to make the transition from thinking only in terms of single cases to thinking in terms of group propensities.

4. Finally, the fourth type of misconception is described as the *causal* misconception where students think there could only be an association when it was attributable to a causal relationship. In the above example, presumably a response might be that there is an association because the personal belief is that smoking causes bronchial disease.

The next stage of the research project was to design computer-based teaching materials to help broaden the students' understanding of the statistical concept of association. The expectation was that computers could extend the statistics taught as well as affect the way in which it was learned by providing students with 'powerful resources and multiple representations' (*ibid.*, p. 226). The detailed reporting of the computer environment produced and the outcomes in terms of students' learning, as well as the access to error analysis that the situation allowed, provide an impressive example of how computers can be used to considerable advantage in this kind of teaching/learning situation. Although there are some reservations about the use of computers in teaching stochastics (Shaugnessy 1992), at this level and for research purposes such as those undertaken in this project, the use of computers appears to be effective.

One of the conclusions to emerge from the study was that students preferred a numerical to a graphical representation of association and they also preferred numerical summaries because of their difficulty in dealing with the idea of distribution. Glencross (1998) suggests that such a preference for numerical representation may be traced to inadequate initial learning experience during which students may not engage in the necessary visual and visualization experiences that are prerequisites to an understanding of graphical representation. In the general summing up of this project, Batanero *et al.* (1998) point to a consequence of their study which relates to 'the distinction between the *personal* (subjective) and *institutional* (mathematical) dimensions of knowledge, meaning and understanding in mathematics' that they found in their students' activities (p. 234, authors' italics). While they gained a personal meaning of association as a result of the teaching programme, they still had not achieved all the intended institutional meaning and still retained some of their earlier misconceptions. The authors state that 'Our results finally show that analysing data is a highly skilled activity even at an exploratory level, requiring a wide knowledge about the problems and concepts underlying graphical, numerical, descriptive and inferential procedures to deal with association' (*ibid.*, p. 235).

The general relevance of this study to mathematics education lies firstly in the message it conveys of the difficulty of understanding statistical concepts even at university level. It would be interesting to know more about the early statistical learning experiences of Spanish children and the extent to which they may or may not have been introduced to the drawing and interpretation of simple graphs and the gathering and presentation of data in simple tables. The results support, however, the conclusions of others such as Shaugnessy (1992), Amir and Williams (1994) and Laridon and Glencross (1994) about the strength of the role of personal belief when it comes to learning stochastics and of the persistence of misconceptions.

IMPLICATIONS FOR TEACHING

This chapter began with references to what and how probability and statistics should be taught in schools. The recommended approach was not to approach stochastics through probability but to ground the teaching and learning of it on the direct experience and experimentation of pupils. However, the research indicates that there is as much concern (possibly more) with probability as there is with statistics based on other types of data gathering. This may suggest a conflict between what is considered relevant and accessible to pupils and what actually goes on in classrooms, but equally it may be that researchers are more drawn to considering probability with its deeper psychological interest. As a result of the imbalance reflected in the research, there are more implications for teaching that relate to probability than to statistical concepts more generally, although some general factors do arise.

Theoretical considerations

1. The strong vein of subjectivism that permeates probability and statistics can make it a difficult subject to teach. Teachers need to be aware of the beliefs, preconceptions and misconceptions their pupils bring to this particular learning situation and to understand how they arise. Pupils themselves will need to come to recognize the interpretive nature of statistics generally, and to come to an understanding of the idea of tendencies as opposed to certainties.
2. The differentiation between causality and conditionality is a difficult one to make, and a confusion between the two affects the judgements made about the probability of events happening. These are difficult concepts for pupils and need to be explicitly addressed.
3. Modelling is advocated by theorists as a meaningful approach to stochastics generally. However, modelling that arises from probability experiments is not considered entirely to fulfil this need with respect to learning stochastics.

Teaching probability and statistics at primary level

1. Statistics brings together several aspects of mathematics and this makes it a useful topic for primary pupils. Children are involved in a practical way in gathering data, representing it and interpreting it. It may be even more meaningful for pupils, however, if the interpretive aspect is developed more, so that through discussion and collaboration, the subjective nature of pupils' interpretations will become more obvious. Misconceptions need to be identified before they can be put right and discussion can play a role in this.
2. The research indicates the importance of being aware of the structures underlying activities we design for pupils and also of what the possible outcomes might be. Some of the evidence points to the heavy reliance on

visual cues in making judgements based on representations of data which can mislead children. Ideas about what is fair in situations like this need to be made explicit in order to help pupils to understand how their misconceptions arise.

3.   When teaching probability, children need to have a variety of devices with which to work, not all with the symmetry of dice, coins and some spinners. When their experience is with this kind of apparatus their ideas of chance become associated with the 'tidiness' of expected outcomes built into them.

4.   The evidence shows that computers clearly can help develop quite sophisticated concepts early on. They can be used in a variety of statistical contexts but they will only be as effective as the program and teacher's intervention allows.

Teaching probability and statistics at secondary level

1.   Again, the message is that learning should be grounded in activity that is meaningful to pupils. Project work is clearly the kind of work that leads to the development of statistical concepts, including data gathering and representation, and the mathematics that follows from both. We have seen how subjectivism persists even beyond secondary years and one way to make this visible and correct misconceptions is for pupils to deal with data that is personally meaningful to them in the sense that they are responsible for it from the point of its interpretation.

2.   The fact that beliefs, superstition and personal experience pervade stochastics brings a multicultural dimension to it that may not always have been obvious. It is clearly important to be sensitive to issues when dealing with the subjectivism with which pupils approach the subject.

3.   Modelling probability situations is an approach that would appear to have considerable potential for developing mathematical concepts. It allows pupils to make conjectures and hypotheses and to test them. In the process of doing so, they are talking about the mathematics they are engaging in, and thinking mathematically in a way which is not always possible at this level.

Chapter 4

Algebra: the Transition from Arithmetic

INTRODUCTION

The earliest stages in the teaching and learning of arithmetic are important to the meaningful learning of algebra. Some of the most basic arithmetic concepts that are taught in the early years of schooling are common to both subjects; the catch is that while they are depicted by the same symbols, they are not always necessarily governed by the same structural rules. Ideas such as equivalence which appear to carry one sort of message at the arithmetic level sometimes are interpreted differently when used in an algebraic context. A more positive example of the links between the two is that when children experience early ideas in learning the concept of fractions, they are being introduced to the 'part-whole' concept which is so fundamental to algebra and the solving of word problems. The rules that apply to solving problems with fractions are essentially the same as many of the rules that apply in work with algebraic equations at higher levels. It is this kind of connection that justifies teaching of the four operations with fractions, for example, when calculators tell us perhaps classroom time could be used more constructively. An awareness of how some of these basic concepts provide the building blocks for mathematics at a higher level can help to identify ways to teach these concepts in the earlier stages, so that they are better understood when pupils meet them at the next level.

Kieran (1992) suggests that algebra is seen by pupils largely as 'memorizing rules and procedures' (p. 390). This is a mechanistic perception of the subject which clearly does not imply the kind of understanding necessary for effective learning; but more importantly, it also suggests that children are not gaining access to a powerful problem-solving tool. In his discussion of Vygostky's activity theory, Mellin-Olsen (1987) says that 'Many of us have intuitively felt that mathematics as a field of knowledge is important knowledge, not only for the benefit of industry and technology, but also for the individual pupils' (p. 47). Algebra is a part of mathematics which is particularly rich in this respect. When a child learns algebra he or she is enriching his or her problem-solving repertoire considerably and although it may seem to be at a level of abstraction that some may consider inaccessible to all children, the simplest modelling procedures used in solving word problems employ basic ideas of algebra that are immensely powerful. The metaphor of a tool is a potent one and its use in connection with mathematics in general and algebra in particular (Sfard 1991, Peck and Jencks 1988) is particularly apt. However, as Streefland (1994) points out, 'From the learner's perspective formal mathematics is distorted mathematics due to its rigidity as a product. This is true for pre-algebra in particular.' Some of the reasons for this will emerge as we examine the nature of algebra and the way in which it is learned and taught.

This chapter will begin with a consideration of what research tells us about the relationship between arithmetic and algebra and some of the problems associated with the transition between the two. The main differences identified by research will be examined and the theoretical perspectives on algebra that help us to understand the nature of these differences will then be considered. This will be followed by considerations of research related firstly to the transition from arithmetic to algebra and then to more specific issues such as solving equations and word problems, symbolization and the use of technology in teaching algebra.

DIFFERENCES BETWEEN ARITHMETIC AND ALGEBRA

It has been suggested that if children can make complete sense of arithmetic and use it as a tool for their thinking, then an understanding of algebra will flow naturally from it (Peck and Jencks 1988). Unfortunately, it is often the 'sense' that children make of arithmetic which, in a way, comes between them and such an understanding of algebra. This anomaly arises as a result of children making rather too direct a transfer of procedures used in an arithmetic context to an algebraic context without being adequately aware of the difference in the constraints that apply at the more abstract level. An analogy drawn by Vygotsky suggests that algebra is to arithmetic what a written language is to speech (Vygotsky 1962). It is not possible to read and write coherently and intelligibly without an understanding of the nature of speech, the words that go to make it up and the general rules that apply to the accepted use of words in conveying meaning. It is not possible to have an understanding of and an ability to 'do' algebra without first of all having an understanding of and an ability to 'do' arithmetic. In algebra, as in language, there are units of meaning such as terms and expressions which have to be used according to particular structural rules which are, in turn, dependent upon the structure of number and the rules that govern their use at the algebraic level. It has been suggested that 'if arithmetic becomes completely sensible to children and becomes a tool for their thinking, the decisions which make algebra sensible flow naturally from it' (Peck and Jencks 1988, p. 85). The emphasis here must be on the words 'completely sensible' as a lack of this depth of understanding at the arithmetic level is a main stumbling block for children's understanding of algebra, as we shall see.

Vergnaud (1997) makes the point that algebra is demanding of children because it not only requires them to carry out symbolic calculations of a new sort but also involves them in learning new concepts such as equations, formulae, function, variable and parameter. Whereas the units of meaning in arithmetic are numbers, the units of meaning in algebra are new mathematical objects, and although the symbols in themselves may carry the same meaning, in algebra they are related to these different mathematical objects (not number only) and different operations (not just addition, subtraction, multiplication and division). Learning to recognize new mathematical objects as entities of a new kind as opposed to a different sort of number is a major difficulty for children in making an effective transition from arithmetic to algebraic activity (Sutherland 1987). This can be seen more simplistically as a jump children are expected to make from the 'reality' of arithmetic to the more abstract world of algebra. At the same time, there is a call for mathematical objects such as those with which algebra is involved to be regarded in a more 'worldly' light. As Restivo (1993) puts it, 'Symbols and notations are simply higher order material objects that we work with the same way and under the same constraints that apply to "hard" material' (p. 257). This is a perspective that has considerable attraction.

Children are able to use intuitive methods when solving arithmetic problems which are directed at finding an answer and avoid using structures implicit in the problem-solving process. However, in algebra the structures cannot be ignored and have to be recognized and used if algebraic problems are to be solved successfully (Kieran 1989, Vergnaud 1997). The research suggests that some of the difficulties children have in recognizing and using these structures arise from the fact that the structures themselves are often disguised by the ways in which children perceive some of the symbols used. They continue to use them as they did in arithmetic and treat them as having a role and meaning identical to that in the arithmetic context without considerations of aspects of structure that cannot be ignored at algebraic level. Without an understanding of this changed perspective of the roles played by some of the symbols with which they are familiar, children will not be able to make the transition as they move from arithmetic to algebraic thought effectively. These differences need to be made explicit if children are to learn algebra in a meaningful way (Chaiklin and Lesgold 1984, Kieran 1992, Boulton-Lewis *et al.* 1997).

Structure in algebra

Structure is a taken-for-granted fundamental aspect of algebra that perhaps needs to be more overtly recognized in teaching. A simple example of structure is given by Freudenthal (1991) when he offers $a + b = c$ as an example of an addition structure. He states that 'Structure is form abstracted from a linguistic expression' (p. 20). In his example, if you add one thing (a) to another (*b*) you get a third (*c*). However, the use of the word 'structure' in the context of algebra is varied and Kieran (1989) has identified some of the complexity in its use.

The simplest view of structure is what Kieran calls *surface structure* which refers simply to the arrangement of the different terms and operations that go to make up an algebraic (or arithmetic) expression. In the expression $2x + y = 7$, the 'structure' is the order in which the symbols making up the expression appear.

Systemic structure refers to the properties of operations within an algebraic expression and the relationships between the terms of the expression that come from within the mathematical system. For example, the expression $5 + 3(x + 2)$ may be written as $3(x + 2) + 5$ (using the commutative law) or $3x + 11$ (using the distributive law to give $3x + 6$, and addition to give 11).

There is also the notion of the *structure of an equation*. This includes within it the structure of the expressions that go to make up an equation (as in systemic structure), but it goes beyond this because it includes the relationship of *equality*. This means that as well as the other kinds of properties included in the systemic structure, the properties of equality (symmetry and transitivity) hold, for example, if the expression in the above example were extended to form one side of an equation as in $3(x + 2) + 5 = 20$ in solving the equation, the balance on both sides of the equal sign must be sustained throughout.

Identifying these levels at which the word 'structure' is applied helps to clarify the source of some of the confusions that arise in children's minds in coming to understand algebra. Throughout the research, the importance of the structural aspects of algebra are referred to frequently. By substituting numbers in the algebraic expressions used in the examples above, it is easy to see how the rules governing arithmetic relationships and operations are inherently the same as those of algebra (Crowley *et al.* 1994). However, it is the importance of making these explicit throughout all stages of teaching number that ensures that these structures become well established and understood by children if they are to learn algebra in a meaningful way (Chaiklin and Lesgold 1984). Two major stumbling blocks for children that emanate from misconceptions about structure are worth pointing out at this stage in order to provide examples of the basic nature of the misunderstanding that can arise. These are the rules that govern the use of the *equals sign* and the concept of an *unknown*.

Children are introduced at a very early age to the use of the equals sign and various symbols to represent an 'unknown'. When faced with an incomplete number sentence such as $4 + 5 = ?$, the message they receive from the equals sign is that something has to be done and that some action needs to be taken with respect to the two numbers (in this case, they have to be added together) (Kieran 1992, Saenz-Ludlow 1998). In the earliest stages children are generally taught to carry out such procedures in connection with physical objects to be counted, usually to answer a question of the 'How many . . . ?' type, but soon they are faced with number sentences with no context (such as that just given) but a result is still expected and, as Kieran suggests, the equals sign is used 'to announce a result'. Rather than thinking in terms of equivalence, children at primary level will often find an answer by working backwards and using a sequence of inverse operations without formalizing either the problem or the solution.

> In arithmetic, the goal is to find the answer, and this goal is usually accomplished by carrying out some sequence of arithmetic operations on either the given numbers of the problem or on the intermediate values derived therefrom. (Kieran 1992, p. 393)

The meaning of equivalence in terms of a symmetric and transitive relationship, so important to algebra, is not necessarily part of children's thinking when faced with simple number sentences of this type, even though it is equally applicable. Instead, the idea of coming up with an answer is paramount in their minds when faced with an equals sign in this context, while in an algebraic problem-solving situation something different is required (Lesh, Post and Behr 1987).

A similar situation arises in connection with the idea of an unknown. Very early in their arithmetic experience children are given questions of the type $4 + \Box = 9$ where the box signifies an unknown number that is 'missing' and they have to find its value. Open sentences of this kind are usually introduced within a context where the numbers relate to specific objects or things. For example, they may be asked how many blocks have to be added to a set of four blocks to give a total of nine blocks. In a situation like this children are able to make an intuitive choice from a variety of strategies such as counting on or counting back (Vergnaud 1997). When the $\Box$ becomes replaced with a letter to represent the unknown as in $4 + a = 9$, children continue to treat the letter a as a label for an object or thing (Kieran 1992). Although this may stand them in good stead with the simplest algebraic equations they meet in the early stages, the situation alters when two unknowns or variables are introduced. When they meet expressions involving more than one unknown they continue to treat them in this way without attending to the structural constraints that are imposed within algebra, so that their method inevitably lets them down.

The difficulties that are built up for children in their algebraic learning as a result of a lack of understanding of the differences in meaning of the equals sign and literal symbols provide examples of the subtle differences of interpretation between the two subjects. Some of these difficulties can be helped at Primary level by simple strategies like stressing the two important principles of equality (symmetry and transitivity). Others have to be made explicit in the early stages of the teaching and learning of algebra to prevent misunderstanding at a fundamental level. The structural principles that govern the use of algebraic terms and expressions account for these differences between the interpretation of symbolic representations in algebra and arithmetic. An examination of some perspectives on the nature of algebra will help to clarify what these structural principles are.

PERSPECTIVES ON THE NATURE OF ALGEBRA

Pimm (1995) describes algebra in the following way:

> Algebra is about form and about transformation. Algebra, right back to
> its origins, seems to be fundamentally dynamic, operating on or
> transforming forms. It is also about equivalence: something is
> preserved despite apparent change. (p. 88)

This statement encapsulates the abstract nature of algebra in spite of an
emphasis on the dynamic aspect of the subject. The notion of 'form' is in itself
abstract and the idea of 'transforming forms' seems to compound this
abstraction. It becomes slightly more solid when Pimm goes on to say 'Algebra
invokes forces that transform: algebraic expressions are shape-shifters' (*ibid.*,
p. 89). The idea of an algebraic expression being a force that can be used to
bring about some change is a useful mental picture that helps to give algebra a
slightly firmer link with the real world.

Procedural and structural aspects

Kieran (1992) gives an overview of the history of algebraic thought which
provides an insight into the nature of the development of the subject. By
considering the nature of what happens as it develops from arithmetic it
becomes possible to identify the fundamental differences between arithmetic
and algebra and thus to gain some insight into how algebra may effectively be
taught and learned. Kieran notes that the essence of the difference between
the two subjects lies in *procedural* and *structural* operations. Arithmetic is seen
to be concerned with *procedural* operations where 'operations are carried out
on numbers to yield n numbers' (*ibid.*, p. 392). A simple purely arithmetic
example is $15 - 7 = 8$. Kieran gives the algebraic example of $3x + y$ where, by
replacing x with 4 and y with 5, the result 17 will be obtained. She makes the
point that although this is an algebraic expression, it produces a numerical
answer which exemplifies the *procedural* nature of algebra. In contrast, the
structural nature of algebra is exemplified by operating on one algebraic
expression to produce another, that is the answer is not numerical but
algebraic. Such an example is $3x + y + 4x = 7x + y$.

The differences between the two lie in the fact that in algebra

(a) the objects that are operated on are algebraic expressions and not
    numerical as in the case with arithmetic;
(b) the operations are not computational as in arithmetic;
(c) the results are other algebraic expressions, not numbers.

Kieran notes that the general approach adopted in teaching algebra begins
with exercises in substitution which reflect the procedural aspect of the subject
and move on to structural aspects of operations such as simplification and

solving equations by formal methods without the transition being made explicit. In doing so, teachers are requiring their pupils to make a considerable cognitive shift without them being made fully aware of the nature of the change. Children are rather abruptly plunged into the more formal and more abstract aspects of the subject without becoming aware of the fact that they are leaving the world of arithmetic with its more tangible foundations behind, and that new rules will apply.

Kieran (1992) identifies the cognitive demands made on children as they progress from arithmetic to algebraic thought. On the one hand, children are introduced to 'treating symbolic representations, which have little or no semantic content, as mathematical objects with processes which do not very often yield numerical solutions' (p. 394). They have to see an algebraic expression which can contain several terms as an entity in itself and which does not relate to any contextual situation. The terms within the expression are not the units; rather, the *expression* of the *way in which the terms are combined* is the unit or object that is to be acted upon. In the expression $2x + y - x$, it is not $2x$ or y or x which is the object but the whole of the expression $2x + y - x$. In being asked to combine terms, they are using a new operation which does not (as in the case of arithmetic) give a numerical answer. The numbers relate to algebraic objects which have no meaning in the 'real' world (even though children tend to treat expressions such as $39x$ as a number so that $39x - 4$ becomes 35).

On the other hand, at the same time children are also 'modifying their former interpretations of certain symbols and beginning to represent the relationships of word problem situations with operations that are often the inverses of those that they used automatically for solving similar problems in arithmetic' (*ibid.*). Kieran gives the following example of a word problem:

> Daniel went to visit his grandmother, who gave him $1.50. Then he bought a book costing $3.20. If he has $2.30 left, how much money did he have before visiting his grandmother? (Kieran 1992, p. 393)

As Vergnaud (1997) and others have established, when solving such problems, children work backwards; for example, in this case they would write $2.30 + 3.20 = 5.50 - 1.50 = 4.00$. The equals sign is taken as meaning 'gives' and does not signify 'is equal to', and it seems to give a left-to-right directional signal to solving the problem. The two characteristics of equality, i.e. transitivity and symmetry, are not respected and there is no attempt to model the problem algebraically.

For understanding of the structural aspects of algebra, the attention of children needs to be drawn explicitly to the structure implicit in problems of this type to help them to realize that their approach is mathematically incorrect. Firstly, the correct meaning of the equals sign must be respected: $2.30 + 3.20$ does not equal $5.50 - 1.50$. Secondly, when written as $x + 1.50 = 3.20 + 2.30$ the transitivity and symmetry of equivalence is preserved. Rather than starting with the final outcome of the problem (i.e. with what Daniel has left) and working backwards as in the arithmetical approach, the total situation is modelled or described in an equation, and *then* a solution is arrived at. It is this sort of apparently simple word problem that often acts as an introduction to the structural aspects of algebra and is meant to lead from procedural approaches adopted in arithmetic to the structural approach of algebra. If such a problem is included within the teaching of algebra but is not solved algebraically, it clearly leads to the perpetuation of applying arithmetic procedures (which are inherently incorrect) in a situation that is intended to be a way into an understanding of algebraic structures.

Process and structure

Sfard (1991) argues that although abstractions within algebra can be conceived of as either structural (objects) or operational (processes), the implicit goals of school algebra are essentially structural, as exemplified by the equation in the problem quoted above. Operational concepts are what children tend to meet first (Kieran refers to these as procedural operations) and tend to dominate the transition from arithmetic to algebra. Although there are structural operations within arithmetic, these are inherently different from those in algebra for reasons which we have already seen (e.g. an arithmetic problem will have a number as an answer). Sfard sees the deep gap between the two as lying in the fact that an algebraic structure has to be seen as a mathematical entity, 'a static structure, existing somewhere in space and time' which has a potential to become 'an actual entity' which is brought into being as a result of a series of actions (Sfard 1991, p. 4). She views the process as 'dynamic, sequential and detailed' (*idem*). However, the fact that it is abstract and divorced from any immediate context is bound to contribute to children's difficulties in grasping such a structure.

This process of viewing algebraic structures as entities (mathematical objects) is the focus for the cognitive leap that children have to make when moving from arithmetic to algebra, and Sfard sees the phases of conceptual development as threefold:

1. *interiorization* – a process is performed on a known mathematical entity (this corresponds with arithmetic procedures);
2. *condensation* – the process is analysed into manageable units (corresponding with the structuring and symbolization of the process);
3. *reification* – something familiar is seen in a new way, and the 'process' becomes 'solidified' into a static structure (corresponding with a 'free-standing' mathematical entity).

These three phases take place over a long period of time and clearly the third phase of reification requires a great leap in terms of the cognitive processes required, and is one which many pupils may not achieve. Thus school algebra is seen as a series of process-object (or procedural-structural) adjustments as children come to understand its structural nature. However, in order to achieve this, it is necessary to recapitulate each of the phases from time to time. There is evidence that children tend to treat arithmetic and algebra as two closed systems, and while it is important to stress building up the arithmetic-to-algebra connection, it is equally important to ensure there is an algebra-to-arithmetic connection to establish that the connection is fully understood (Lee and Wheeler 1989). Children should become accustomed to working back and forth between algebra and arithmetic and be able to identify the advantages of each, depending on the nature of the task with which they have to deal. It is considered that in this cyclical interchange between the two, a better understanding of the structural nature of algebra will emerge. This highlights the importance of making the connection between natural, intuitive reasoning and formal calculations as they arise in the first instance in arithmetic (Vergnaud 1997), and of helping children to build appropriately on the intuitive knowledge they bring to mathematical problem-solving situations (Fischbein 1987) so that they can learn algebra meaningfully.

We shall now go on to consider how these perspectives on algebra inform the research related to the classroom teaching and learning of algebra.

PRE-ALGEBRA: FROM ARITHMETIC TO ALGEBRA

Lee and Wheeler (1989) see 'the track from arithmetic to algebra to be littered with procedural, linguistic, conceptual and epistemological obstacles' (p. 57). There is a considerable body of research focusing on the transition from arithmetic to algebra, some of it concerned with highly specific stumbling blocks, such as the use of the equals sign as we have already seen, and some with the differences in approach such as those in solving word problems in each subject. Much of the teaching of algebra to children can be treated as a transition period bridging what Gray and Tall (1994) have called the 'cognitive gap' between the two subjects. As we shall see, children tend to carry with them the perspectives and processes established in arithmetic to fall back on when, in later stages of schooling, they are faced with algebraic situations which they cannot understand. The notion of 'pre-algebra' has come to be fairly well established in the research, which helps to draw attention to the importance of this phase of teaching and learning the subject.

Literal terms

Hersovics and Linchevski (1994) suggest that the cognitive gap between arithmetic and algebra 'can be characterized as *the student's inability to operate spontaneously with or on the unknown*' (p. 59, authors' italics). However, some of the source of difficulty using literal terms must also lie in the differences in what they may be depicting and the ways in which they are used. Harper (1987) draws attention to this kind of confusion when he notes that children are taught to view literal labels as unknowns in the earliest stages of learning arithmetic and solving simple word problems, but then move from this very quickly to the notion of a letter as a 'given' when calculating a set of values from mappings. A common situation is for children to be presented with their multiplication tables in the form of mappings, for example, as shown in Figure 4.1.

x	×3 →	Answer
1	→	3
2	→	6
3	→	9
4	→	12

Figure 4.1

Harper suggests that there are stages through which children have to progress in order to understand a literal term as a variable and that very often they are expected to be able to conceptualize it in this way before they can actually see the general in the particular. This means that children acquire procedural conceptions of literal terms before they understand the structural aspects of algebra that govern them, and the implication is that the procedural conception dominates unless the structural conception is made explicit. They are being asked to express the idea of a numerical pattern in an algebraic way before they fully understand the concept of generalization and possibly before they have a meaningful understanding of the concept of a numerical pattern.

The use of the equals sign

The pivotal importance of the equals sign and how it is interpreted was noted earlier, and this importance is reflected in the research. One such study was carried out by Saenz-Ludlow and Waldgrave (1998) who investigated the interpretation by Grade 3 pupils of the concept of equality and the use of the equals sign. The children were taped in the course of lessons and the transcripts were analysed to determine whether there was a pattern in the way in which children's understanding of the concept developed. It was found that children's interpretation of equality changes and develops gradually. Initially, the equals sign is taken to be a command which indicates that the numbers given have to be acted upon, and only later did the children make the connection between 'quantitative sameness on both sides' of the symbol (*ibid.*, p. 185). The role of discussion among the children was found to play an important part in developing their understanding in progressing to this new interpretation. The children's dialogue about performance on the tasks they were given indicated the strength with which they could defend their interpretations and their approach to doing tasks. Underlying all of this was the commitment on the part of the teacher to listen to the children, to adapt to the current level of their knowledge, help to identify their difficulties and so help them to progress in their understanding of the meaning and function of the equals sign.

Dialogue between children and children, and children and teacher, is also identified as a vital aspect of children's learning of symbolization processes in a study carried out by Saenz-Ludlow and Waldgrave (1998). They found that children's dialogue, both with each other and with the teacher, 'raise our awareness of the cognitive effort entailed in the interpretation of and the construction of mathematical meanings from the conventional symbols' (*ibid.*, p. 185). They suggest that there are two complementary procedures going on: on the one hand, children are having to construct arithmetical meanings for equality, and on the other, they are having to learn to use the symbolic language involved. In the course of this 'symbolic activity', children write their 'spoken symbolization' which is seen as 'a transitional step in expressing in conventional symbols' what they have done, without losing what those symbols are meant to signify (*idem*). In the process of expressing the meaning of the equals sign in their own words and working from this through dialogue with each other and with their teacher, they come to recognize and understand the factors that are essential to the correct interpretation of the equals sign, reinforcing the conclusions of studies reported earlier (Kieran 1992, Sfard 1991, Lee and Wheeler 1989). The importance of verbalization in coming to understand symbolic meaning in the transition from the arithmetic to the algebraic level is further evidence of the general importance of language in constructing meaning for mathematical concepts generally, as noted throughout this book. Discussion is seen as important in a constructivist approach to teaching and learning in the classroom as a means of reaching a shared understanding, but it also provides teachers and researchers with an insight into how pupils reach this understanding (Goodchild 1996).

Cognitive obstacles

In an extension of earlier work, Linchevski and Herscovics (1994) explored the nature of cognitive obstacles in children's pre-algebraic thinking. A rationale for carrying out such work is given which pinpoints the need to make a systematic demarcation between the first courses taught in secondary school algebra and pre-algebraic thought in the hope that the necessary factors that characterize this transition will be treated more explicitly in the classroom. The study involved Grade 7 pupils (approximately 12 to 13 years old) and had two aims:

(a) to try to evaluate the pre-algebra potential of younger pupils;
(b) to expand the assessment made of pre-algebraic arithmetic skills to see whether the cognitive obstacles already identified might exist in a purely arithmetic context, and if they did, to find out how widespread they were.

The children in the sample had already been taught the order of operations. Five obstacles were identified as follows:

1. *Over-generalization of the order of operations.* Pupils in the sample were asked to evaluate the following strings:

    a) $5 + 6 \times 10 = ?$
    b) $7 - 3 \times 5 = ?$
    c) $27 - 5 + 3 = ?$
    d) $24:3 \times 2 = ?$
    (Linchevski and Herscovics 1994, p. 177)

    They found that pupils tended to see addition as having priority over subtraction. Where subtraction was involved, they tended to split the expression with the minus sign and proceed from there. Only Question (c) was solved correctly by the whole sample.
2. *Failure to perceive cancellation.* This focuses on the way in which children continue to work from left to right when dealing with expressions of this kind and do not look at the whole to determine where they might cancel terms or gather them together. For example, only 59.3% of the sample obtained the correct answer for $329 + 167 - 167$.
3. *A static view of the use of brackets.* Pupils in the sample understood that brackets meant 'do this first' and the whole sample were successful with $8x$ $(5 + 7)$. However, once the context changed and they were asked to say whether $926 - 167 - 167$ was the same as $926 - (167 + 167)$ only two of the 27 pupils thought this was the case.
4. *Detachment of a term from the indicated operation.* In earlier work, Linchevski and Hercovics had found some evidence that the pupils tended to detach a term when subtraction was indicated. For example, in $27 - 5 + 3$, they would perform $27 - 5$ first, then add the 3. There was some variation as to whether they did this in all strings, and their explanations for doing so included statements such as '. . . what's convenient is the basis for my decision. If they were mixed with multiplication, I would have to be careful' (*ibid.,* p. 180).
5. *Jumping off with the posterior operation.* This obstacle refers to the way in which pupils tend to group like terms where a distance between the terms is involved and focus on the operation following the term. Given the string $217 + 175 - 217 + 175 + 67$ only 70.3% cancelled the 217 correctly and focused on the minus sign after the second term. Two pupils actually refused to do these tasks, saying that 'you have to go from left to right, I can't just jump' (*ibid.,* p. 181).

Following on the first part dealing with arithmetic obstacles in pre-algebra, pupils were then given linear equations to solve. Most students were able to solve the equations but 'using procedures based exclusively on numerical manipulations or numerical substitution' (*idem*). Where they failed to solve the equations successfully, the researchers were able to trace the failure to some of the cognitive obstacles just described.

This study is particularly important in the light of work reported earlier in connection with the cognitive gap between arithmetic and algebra (Gray and Tall 1994). It draws attention to specific issues that can be addressed in the classroom at the pre-algebra level. Linchevski and Herscovics (1994) advocate making the order of operations clearer to pupils so that they are seen as necessary and not arbitrary. The introduction of brackets and their operational use also needs to be dealt with more explicitly and preferably through a new means of presentation, as does the tendency of pupils to over-generalize. This could be an instance where the calculator comes into its own more strongly and where, in using it in an exploratory way, pupils are able to see quite quickly the different outcomes that occur when the rules are not adhered to.

Boulton-Lewis *et al.* (1997) investigated the transition from arithmetic to algebra in a longitudinal study involving the introduction of the concepts of variable, equation and solution of linear equations to pupils of Grade 8. Concrete materials (manipulatives) were used as part of the teaching approach, and the outcome of this part of the study was quite unequivocal since, after initial instruction in which they were used, not one of the 21 pupils in the class used the concrete materials to help them in the process of solving linear equations. A majority (fourteen) 'used inverse operations in the reverse order' thus supporting other evidence reporting the tendency of children to apply arithmetic processes to the solution of algebraic problems (Sfard 1989, Kieran 1992, Linchevski and Herscovics 1996). The conclusions of the study were stated as follows:

> Firstly, the inverse operations strategy uses a different conception of equation (two-way change) from the taught approach (balance). Second, the students' responses reflect the findings of Hart (1981) in that a gap exists between concrete and symbolic representation. Third, the finding seems to reinforce the heavy cognitive load involved in using containers and objects. Fourth, the finding also seems to support Kieran's (1992) argument that algebra knowledge develops from procedural to structural as inverse operations is a procedural strategy while the containers and objects approach seems to have structural tendencies. (Boulton-Lewis *et al.* 1997, p. 190)

The need for pupils to understand an algebraic equation as a sequence of operations was also highlighted.

Part of this study involved interviewing pupils in Grade 7 to gain some knowledge of what they knew about operations, operational laws, equals and variables. It was acknowledged that 'The students would have been taught about operations and equals as part of the curriculum in arithmetic, but any knowledge of variables and linear equations could only have been derived intuitively from their knowledge of arithmetic' (Boulton-Lewis *et al.* 1997, pp. 190–1). From the questions asked, it was found that only about 26% of the sample were able to identify the correct order of arithmetical operations and then to apply them in the context of solving linear equations. Once again it was found that 'With regard to the equals sign in an unfinished equation with a series of operations, almost 100% of the students believed it meant find the answer, and in the completed equation only half of the students could say that it meant that both sides of the equation were the same' (*idem*). At least half did not have the concept of equivalence. This substantiates further the evidence that, faced with this type of algebraic problem, many children would take the meaning of the equals sign to be that they 'have to find an answer'. One of the implications drawn from this study was that most of the children involved needed a 'better understanding of division and the order of operations in complex arithmetic' (*ibid.*, p. 191). Another important factor identified by Boulton-Lewis *et al.* was that the children in their study did not understand the difference between the use of x as a variable and x as an unknown when solving equations that involved a multiple of the variable. For example, they viewed $3x$ as a variable in itself rather than as a multiple of x.

The difficulties the children faced in the above study clearly indicate a lack of knowledge of the structural rules that apply in algebra. Sfard and Linchevski (1994) note this kind of difficulty and the difference children have in transferring operational or procedural rules from arithmetic to algebra, compared with the transference of structural rules. They take as an example the expression $3 (x + 1) + 1$ and state that 'Although semantically empty, the expression may still be manipulated and combine with other expressions of the same type, according to certain well defined rules' and children handle these relatively easily (*ibid.*, pp. 87–8). They note that the operational approach to formulae is inherited from arithmetic and they argue that:

> Spontaneity of the operational outlook is exhibited in the easiness with which many young children handle simple linear equations of the form $ax + b = c$. To them it is often 'intuitively obvious' that they simply have to 'undo' the equation to arrive at the unknown. Whereas this intuitive understanding and adoption of an operational outlook appears to come into play at this level, the same intuitive understanding or acceptance is not in evidence with deeper aspects of the structural approach. Whereas pupils can cope with the equation $7x + 157 = 248$, they are not able to cope with $112 = 12x + 247$. The simple matter of having the unknown on the right confuses them: 'It's not like it should be'. (Sfard and Linchevski 1994, p. 209)

An example of this kind of confusion is given by Booth (1984). She found that while primary schoolchildren understood that the total number of terms in two sets is 13, they could not recognize that the set of 13 could just as well be represented by $8 + 5$, for example. If they cannot perceive it in this way, they are, as a result, unlikely to be able to recognize $a + b$ as an algebraic object and an entity in itself (see below). There is within arithmetic 'an intuitive precursor' of $a + b$ as an algebraic object, but when it continues to be treated without the structural constraints and rules that apply, then children have difficulty (Kieran 1989). They have to understand, for example, that $a + b - c$ is not equal to or the same as $a - b + c$. The syntactical rules governing algebraic equations have to be obeyed.

SOLVING PROBLEMS

Commenting on two different pupils' approach to algebraic problem-solving, Sfard and Linchevski (1994) state that:

> It seems, however, that the meaningfulness (or should we say meaninglessness?) of the learning is to a great extent, a function of students' expectations and aims: the *interpreters* will struggle for meaning whether we help them or not, whereas *doers* will always rush to do things rather than think about them. The problem with the doers does not stem so much from the fact that they are not able to find meaning, as from their lack of urge to look for it. (p. 264, emphasis added)

The meaninglessness of the problems that pupils are generally faced with in algebra are a cause of concern for many researchers (e.g. Douady 1997, Davis 1988, Noddings 1993, Sfard and Linchevski 1994). The problems used in most teaching and textbooks do not invite pupils' involvement essentially because they do not arise from any of the pupils' activity and they cannot relate them to anything remotely tangible in their lives; thus they have no ownership over them. The problems usually remain meaningless, however motivated pupils may be, and yet problems should be the pupil's point of entry to the algebra being taught. In writing of a problem- or project-oriented approach to mathematics generally, Skovsmose (1994) answers his own question, 'Why problem orientation?' with 'Because a specific problem can become the point of entry to a complexity; a totality can be made comprehensible by an intensive study of a central problem' (*ibid.*, p. 78). In most algebra classrooms, this appears not to happen.

There are two kinds of problems with which pupils are typically faced in learning algebra in the classroom. The first are straightforward algebraic equations to be solved, without any context other than the mathematical one. The second kind is the 'word problem' where the problem is posed within a context and involves the use of algebra; firstly, in modelling the problem, then expressing it as an equation, and finally solving it. Word problems are well known at the arithmetic level and the distinction between what is known as a word problem in arithmetic and an algebraic word problem is difficult to define, if indeed it exists (Kieran 1989).

Solving equations

When solving equations with a single unknown, as we have seen, children can fall back on their knowledge of arithmetic and find a solution using a variety of strategies. However, when more than a single variable is involved these strategies let them down. They are faced with a situation where they have to do something *to* the equations without any contextual clues as to how to proceed, particularly if they do not fully understand the structural rules that govern the equals sign and its use. Pimm (1995) suggests that 'sequences of equations show the results of transformations, not the transformations themselves. Hence, algebra takes place between the successive written statements and is not the statement themselves' (p. 89). This clearly suggests that (unless graphing is carried out as they proceed) something is done *to* equations (whether a single one or a series) and not *with* them, and while equations are the product, algebra is the process.

Chaiklin and Lesgold (1984) attempted to find out the extent to which children of approximately 11 years old had a satisfactory knowledge of algebraic structure by asking them to judge the equivalence of several numerical expressions without actually computing the total. An example they used is as follows:

$$685 - 492 + 947 \quad 947 + 492 - 685$$
$$947 - 685 + 492 \quad 947 - 492 + 685$$

They found that pupils used several different methods to combine the numbers, even within a single expression. It was evident that they relied on what they thought the arithmetic outcomes would be and preferred to calculate rather than judge equivalence on the basis of either number principles or the rules that govern the manipulation of symbols. Those that did try to make a judgement based on algebraic rules of transformation had difficulties which often produced the wrong result.

Technically, solving an equation involves performing the same operation on both sides of the equation. Evidence suggests that the first thing pupils do is to guess first and then test their guess. Kieran (1992) lists the various strategies used by children that have been identified:

1. The use of number facts (e.g. $3 + n = 5$; $5 - 3 = 2$; $\therefore n = 2$).
2. Use of counting techniques (e.g. $3 + n = 5$ and they count on from 3 to 5 to find $n = 2$).
3. Cover-up (e.g. $2x + 9 = 5x$; 9 *must equal* $3x$; $\therefore x = 3$).
4. Undoing or working backwards (e.g. $2x + 4 = 18$; $18 = 2x + 4$; $18 - 4 = 2x$; $16 = 2x$; $x = 8$).
5. Trial and error substitution.
6. Transforming (changing sides and changing signs).
7. Performing the same operation on both sides.

The last two techniques are formal techniques in that they involve knowledge and application of algebraic structural properties, whereas the first two are arithmetical in nature and the third and fourth might be seen to be a combination of both. The research suggests that those pupils who have been taught both by the 'cover-up method' or to perform the same operation on both sides of an equation are most successful, while those who have been taught to solve them formally only were less successful.

When given open-ended, generalized problems, Harper (1987) plots the progression of pupils' ability to handle symbolization in algebra. He used the rhetorical method in which statements such as the following were presented:

> If you are given the sum and differences of any two numbers, show that you can always find out what the numbers are.

He found that in a sample of 144 pupils in secondary school, in the first year, children begin by answering the question using words. By second and third years, they produce a solution that involves an unknown and represent the equation as a problem to be solved. By the third year, some begin to move on from the representation of the situation as an equation and he found that about 20% of a sample of children were able to give the answer in the form of a generalization, in this case $m + n/2$ and $m - n/2$.

Harper argues that this progression through approaches to representing a problem in words first of all and gradually reaching a generalization models the general development of this cognitive skill and does not merely reflect teaching style. If this is the case, it should then *inform* teaching style. Douady (1997) reinforces this point when she says that '*Writing* and *solving* equations are *essential working tools* for modelling questions and answers about the relations between magnitudes' (p. 384, author's emphasis).

Schoenfeld (1987) gives a description of the tortuous procedures through which some pupils go when solving equations. He gives the following example from a videotaped session in which pupils were given the equation $x^2y + 2xy + x - y = 2$:

> They began by moving the lone y over to the right-hand side of the equation, which seems reasonable. They brought together the y terms on the left-hand side of the equation, factored the common y, and then factored the common x term from the expression as well. Then they moved some terms over to the right-hand side . . . and so on. Each manipulation was correct. But a close look at what they did reveals the equation they were working on after six algebraic steps was more complicated than the one they had started with! They simply proceeded one step at a time. Each time they got a new equation they seemed to start over: 'Now what can we do with this equation?' As a result, they dug themselves into a deeper and deeper hole, without standing back to see if what they were doing made sense. (Schoenfeld 1987, p. 192, author's italics)

This is a familiar classroom scenario and Schoenfeld uses this example to establish the importance of what he calls 'self-regulation' in problem-solving: 'It's not only what you know, but how you use it (if at all) that matters' (*ibid.*).

Children form their own perceptions of what is happening in the early stages of solving equations in a single unknown, which, if allowed to go unaddressed, will let them down in later stages. Such an example is offered by Soh (1994) where a child was solving $3x + 15 = 5x$. He said, 'Can't do that . . . unless you switch them round.' He believed that to solve an equation with one unknown, there had to be more of it on the left-hand side. Sfard and Linchevski (1994) refer to this kind of difficulty when discussing the way in which children invoke arithmetic structures when solving equations. They note that to the children it is often 'intuitively obvious' that they simply have to 'undo' a linear equation such as $7x + 157 = 248$ whereas they cannot cope with $112 = 12x + 247$. The order of the equation confuses them and 'it's not like it should be' (p. 209). Once again it is the structural aspect underlying the use of the equals sign that eludes them.

Particular concepts such as parameters seem to pose great difficulty for pupils. Bloedy-Vinner (1994) suggests this is in part due to a deficiency in the understanding of algebraic language which he found a source of confusion with high school pupils. Drawing on the work of Kuchemann (1981) and Usiskin (1988) the main use of letters is described in terms of generalized number, specific unknowns in equations, and variables. However, letters are also used to denote parameters, as in the example of the graph of the linear equation $y = ax + b$. Confusion arises because no indication is given as to which letters represent parameters and which are variables or unknowns. Also, as in this case, the role of the parameters a and b changes; where they can begin as the unknown they become the 'knowns' and a final equation in solving a problem of this type would be $y = 3x - 1$.

Lins (1994) focuses on epistemological considerations in studying the relationship between the theoretical construct of the *semantic field* and *solution-driven* and *justification-driven* activities in algebra. His work is an example of new perspectives within research in mathematics education and its practical relevance may not at first seem clear. However, new perspectives of this kind can be powerful in shifting our thinking in a way that causes us to focus on what may appear to be the obvious; but because it is so, it may often be ignored. Lins draws on Davydov's (1962) work which was touched on earlier in connection with word problems, and he also acknowledges the influence of Vygotsky.

Lins (1994) states that 'At the heart of TMSF [Theoretical Model of the Semantic Fields] is a particular conception of knowledge: *knowledge* is a pair formed by a *statement-belief* and a *justification* for it' (Lins 1994, p. 185). He gives the example of the equation $3x + 10 = 100$ and suggests that in solving it, one might decide to take 10 from each side, which is a *statement-belief*, and justify doing so by saying that it is like a situation involving balance in a pair of scales *(justification)*. He points out that this justification would not hold if the equation were $3x + 100 = 10$. This would require a different justification with a resultant product of new knowledge, because taking 10 from each side produces a new expression that does not resemble a 'balance', i.e. $3x + 90 = 0$.

In the TMSF, Lins points out that *meaning* is seen as the relationship between the statement-belief and the justification. Taken together, these are seen as the two elements of knowledge, and in order for some part of mathematics to be meaningful to a person, that person must have knowledge of it; they must be able to manifest a statement belief in some way and be able to justify it. Meaning is produced through, or within, a semantic field. Lins uses the equation above as an example to explain the meaning of a semantic field:

> A *Semantic Field is a mode of producing meaning.* We can speak for example, of producing *meaning* for the equation *3x + 10 = 100* within a *Semantic Field of a scale-balance*, or within the *Semantic Field of algebraic thinking* ... or within a *Semantic Field of whole and parts.* But within the first or the last of these *Semantic Fields*, it is not possible to produce meaning for the equation $3x + 10 = 100$. (*ibid.,* author's italics)

Only when $3x + 10 = 100$ is recognized as an algebraic expression does it have meaning.

Lins used the concept of semantic field in research carried out over a series of lessons with 11–12-year-olds in a Brazilian school. They were presented with the problem given below (similar to that used by Davydov, see Chapter 1) and asked to work at it through small-group discussion.

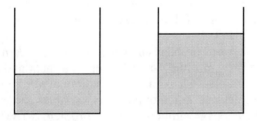

With nine more buckets, the tank on the left will be full; with five more buckets, the tank on the right will be full. What can one say about this tank situation? (Lins 1994, p. 187)

Figure 4.2

They generated a series of expressions of their own to depict what they found about the situations they had chosen and in each case had to justify these. For the next stage, they had to propose a different approach to the justification for their expressions. For example, they were asked to find a transformation for one of the expressions, $x + 9b = y + 5b$, which would lead to one of the other expressions, $x + 4b = y$. Because they had derived the expressions themselves in the first place, they apparently had no trouble in performing such a transformation, for example by saying *take 5b from each side*. Lins refers to this as a new mode of producing meaning and points to the twofold importance of pupils having to give a justification: first, they supported the manipulation of the expressions as one way of making sense of producing new expressions; second, by using activities that focused on justifications (as opposed to finding a numerical solution), the activities had to make sense to the pupils. They were put in a position of having to think and act for themselves rather than producing an answer that might be seen merely as 'correct'. In Lins' words, 'once the direct manipulation of expressions had become a "senseful" activity, not only the technical difficulties did not occur, but also the students began to bring into play methods produced in arithmetic (for example, simplifying simple rational expressions)' (*ibid.*, p. 189).

A very important feature of this research lies in how it illustrates the advantage of having pupils generate their own equations from a fairly realistic situation which they can at least identify, if not one to which they can actually relate. They understood the action of balancing, for example, by subtracting $5b$ from each side of the equation which they had derived, since mentally they could link the 5 with a number of buckets. The meaninglessness is reduced and they have a problem situation for which they have generated their own symbolic representation, and this allows them to take these forward to solve the problem which has some meaning – at least, for them.

Word problems

Noddings (1993) suggests that mathematics teachers 'are baffled by the seemingly Herculean task of locating themes, basic interests, or occupations that might serve as springboards for the systematic study of algebra, geometry, or trigonometry' (p. 153).

However simple word problems may appear to be, it seems that children find it difficult to recognize the similarity between them and specifying the relations among variables (Chaiklin 1989). Traditional word problems tend to lead pupils to a direct translation of numbers, variables and constants in the order in which they appear within the written problem (Crowley *et al.* 1994). What seem to be small differences between problems can have a strong effect on children's ability to solve them successfully. Adopting a more generic approach to solving such problems which entails the same important factors of being able to analyse a problem, to isolate the specific question being asked and to recognize the main features essential to its solution, has been tried in Russia by Davydov (1962) and reported extensively by Freudenthal (1974).

In many countries, the approach adopted in teaching algebraic (or indeed arithmetic) problem-solving is to ask children first of all to formulate an equation containing unknowns and specifying operations, next to manipulate the terms of the equation, then isolate the unknown, and finally, to find its value. In complete contrast to this, Davydov developed an apparently effective approach to solving problems algebraically with Russian children of 8 years old onwards which entailed a method that focuses on the part-whole aspect of word problems. The salient points (the parts needed to answer the question being posed) are embedded in a context (the whole) and the main thrust of the methodology is to develop children's understanding of the part-whole relationship through a variety of experiences. To begin with, concrete materials are used: for example, strips of paper cut into parts, a collection of materials of different weights, or containers with different volumes. The purpose is to establish in as concrete and overt a way as possible what is meant by 'part' and what is meant by 'whole'. The children record the results of these activities with drawings or letters, but never with numbers, and the relationships of the whole with its parts are depicted using the symbols +, – and =. A considerable amount of time over several weeks is spent on performing operations of this kind using the drawings, letters and signs, during which children are engaged in writing 'formulae' to represent the situations. An example offered by Freudenthal is given below:

> The teacher shows a graduated glass where the water level (k) is marked by an elastic band. Then he takes another graduated glass with c water and pours both into a drinking glass, which now contains b water. The pupils make a drawing, representing the connection between the magnitudes k, c, b and note down the formulas $c = b - k$, $b = c + k, k = b - c$. (Freudenthal 1974, p. 399)

The pupils are expected to write down all of the possible part-whole expressions and to choose the one that describes the situation they have just observed.

During the next stage, children are given written word problems which they have to represent first of all in a drawing which they then translate into a schema and present in formulae. Finally, they have to write texts to match a situation presented in a drawing and arrive at a formula. Freudenthal sees this as 'unpacking' the problem where the whole or 'principal dependency' is identified first of all and the subsidiary components successively identified (*idem*). The approach has been shown to be effective, with 8-year-olds being able to model problems using literal terms only, and continued to be effective with 9- to 11-year-old children using progressively more difficult problem situations. The approach clearly builds on the main concepts within word problems as being essentially part and whole, followed by the careful sequencing of stages in presenting these concepts in different ways, and finally modelling them. It reminds us once more of the importance of the concept of fractions in algebraic problem-solving. (One need not go much further than to recognize the fact that almost all word problems in the TIMSS study are classified under ratio and proportion.)

Lesh (1987) describes the solution of a word problem as having three significant translations:

1. from English sentences to an algebraic sentence;
2. from an algebraic sentence to an arithmetic sentence;
3. from an arithmetic sentence back into the original problem situation.
   (p. 198)

The example below could illustrate what he calls the translation-transformation procedures that constitute the process of modelling in algebra.

> Nada has 15 apples. This is three times the number of apples Kevin has. How many apples does Kevin have? (1) Let x be the number of apples Kevin has. Then $3x = 15$. (2) If 3 times a number is 15, then that number must be $15/3 = 5$. (3) Kevin has 5 apples.

This is an example in which the literal term is an unknown, and clearly the translation-transformation becomes more difficult when it becomes a variable. This is borne out by Kaput (1987). He notes that erroneous equations are often produced when reasonably sophisticated students are faced with translating from tables of data or diagrams into an algebraic symbol system. This is a situation which demands n additional mediating steps whereby the quantitative relationship is first translated into natural language. He notes that 'data involving mathematically mature subjects indicate that the natural language encoding difficulties are present even in sophisticated subjects', and to produce correct equations, they often have to overcome 'the natural language "default" encoding process that leads to error' (*ibid.*, p. 187). The other main factor that leads to the construction of incorrect equations is the 'students' weak understanding of the notion of the algebraic variable and hence the weak alternative to the strongly learned natural language encoding process to which it falls victim' (*ibid.*). This suggests that the poor understanding of the concept of variable is persistent at higher levels of study. The importance of language in formulating algebraic equations has also been identified by Arzarello *et al.* (1994).

Bednarz and Dufour-Janvier (1994) have carried out an analysis of word problems presented in arithmetic and algebra textbooks at different grade levels in Canada covering 12- to 16-year-olds. They selected questions to be used with groups of pupils from those who had had no experience of algebra to those who were dealing with equations of several variables. They report several criteria of complexity including the effects of the part-whole aspect of a problem, the influence of sequencing relationships and the number of relationships involved.

A study of pupils' ability to write equations to represent word problems with a single unknown has been carried out by Cortes (1998) working with Grade 8 classes in French schools. She classified these word problems into three types and she states:

> It is possible to construct a single-unknown equation by: I – substituting unknowns with given numbers and units into a given formula or into a constructed function. II – substituting unknowns with linear functions into a two (or more) unknown equation. III – equating two linear functions. (*Ibid.*, p. 210)

She argues that pupils engaged in solving word problems are often using the notion of a numerical function and therefore have an 'implicit and pragmatic concept' of a function. Sub-categories of problems within the three mentioned above were identified and the children's performance on them recorded. She found that there was little problem with syntactic translation identified by MacGregor and Stacey (1996) but there were frequent 'analgebraic' errors (Bloedy-Vinner 1994) when *x* and *y* were given the same status in systems of equations. It was also noted that students reverted to arithmetic when the problems were either too easy or too difficult. The more data given in a problem, the more errors there tended to be and she suggests that this is due to the fact that pupils are inclined to 'use all the data' as a rule of action in solving such problems. This may be due to cognitive overload or to other factors such as the bad wording of problems. In conclusion, Cortes notes that the concept of function comes into the curriculum *after* the introduction of word problems and agrees with Yerushalmy (1997) in suggesting that it should form a major focus for the algebra curriculum rather than basing it on the performance of mechanical manipulations. This is not surprising since her classification of word problems suggests that the notion of a function is present in many of them. It is also noted that the classification of word problems and models of pupils' processes (not presented here) have been helpful to teachers in the analysis and treatment of pupil errors.

Modelling

Lamon (1998) adopts the view of Kaput (1993) in research related to how the algebra curriculum should be approached through focusing on understanding quantitative relationships and mathematizing authentic experiences. Modelling is seen to provide a powerful means of focusing pupils' attention on the variables and parameters that are inherent in problems and identifying the relationships between them. She notes that:

> Focusing on the reasoning that precedes the production of a symbolic equation provides students the opportunity to see algebra as an activity; to appreciate its utility; to see familiar situations as a source of meaning for formal mathematical symbols; to develop a systematic and firm foundation for more abstract reasoning. (p. 168)

A three-course pre-algebra and algebra course was devised for Grade 7, 8 and 9 pupils. Pupils were presented with a model for analysing the problems given them.

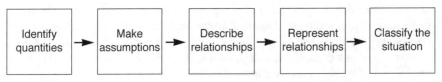

Figure 4.3

The pupils 'analysed and represented quantitative relationships in verbal, pictorial, graphical and tabular form' (p. 170). Although mathematically able, they had difficulty in verbalizing and writing about the mathematics they had engaged in. By the end of their first year 'the algebraic language, concepts, meanings, and reasoning that they had acquired, were well connected, practical, and grounded in familiar situations'. Having taken the same test as a Grade 9 class who were not involved in this experimental approach, the Grade 7 pupils compared favourably with them on their performance. The Grade 9 pupils 'could not reason quantitatively nor solve simple modelling problems' and although they had finished their formal study of algebra, they had done so 'with lots of tools and no sense of when to use them' (p. 174). Again, we are reminded here of the potential algebra holds as a personal resource and problem-solving tool for all pupils (Mellin-Olsen 1987, Sfard 1987, Kieran 1992).

The study by Lamon uses something of a template approach to guiding pupils through the processes in problem-solving. Another approach to modelling is taken by Lemut and Greco (1998). They consider that algebraic modelling is based on three major components:

1. systemic thinking;
2. mastery of a range of representation systems with rules for manipulating the elements of those systems;
3. the capability of converting representation systems to different representation systems.

Systemic thinking is defined as *'thinking globally but acting locally'* (*ibid.*, p. 192, authors' italics) and is seen as involving modelling and controlling manipulations within an algebraic context. They conclude that it is important to engage pupils in modelling to help them to 'make explicit the relations among the involved entities' and that modelling and manipulation should be treated separately (*idem*). The modelling helps pupils to focus on the structural aspects of a problem, while manipulations helps to make the solution that is implicit in the model more explicit. This is reminiscent of Davydov's (1962) part-whole approach.

Gender differences in algebraic problem-solving

Kota and Thomas (1997) carried out a study in New Zealand to investigate affective factors in children's ability in solving algebraic problems, to analyse any gender differences and to explore reasons for them. The sample consisted of 345 children aged 13 to 15 years with single-sex and co-educational schools both represented. A questionnaire was given to the children to complete at the beginning of the study, to measure affective factors, as well as a short test consisting of five basic word problems which had been previously piloted. The affective factors comprised self-concept, interest, anxiety, self-perception, usefulness of mathematics, mathematics-intrinsic motivation and enjoyment of mathematics.

An example of one of the five questions is given below:

> A wooden fence is made by placing 3 planks between two posts as shown in the figure. (A picture was given.) Complete the table to find the number of planks needed for fences of 5 posts; 100 posts and S posts. Write an algebraic equation to find the number of planks L in a fence S posts long.

Number of posts   2   3   4 5   100 S
Number of planks 3   6   9

(Kota and Thomas 1997, p. 155)

While the data given relate to two different groups (one from Form 3 and one from Form 4) they were from the same population. Before taking gender into account, they found that the mean score for all affective factors except the usefulness of mathematics decreased from Form 3 to Form 4 and was particularly significant for self-concept, anxiety and enjoyment of mathematics. At the same time, there was also a significant increase in the ability to solve algebraic problems. When the data were analysed according to gender, it emerged that in Form 3 the mean scores on all affective factors were higher for girls than for boys, except mathematics–self-perception and enjoyment; however, this situation was reversed in Form 4. Girls performed significantly higher than boys in algebraic problem-solving in Form 3, and although still higher than boys in Form 4, it was not significantly so.

A different pattern emerged when the scores of the two groups of girls were compared with each other. The mean scores for four of the affective factors declined significantly for girls; these were self-concept, interest, anxiety and enjoyment. On the other hand, the mean scores for the boys increased from Form 3 to Form 4 on all affective factors, although not significantly. The girls' mean problem-solving score increased from Form 3 to 4, but not significantly, and significantly less than the increase by the boys.

It was found that where boys' mean scores decreased, they did so at a slower rate than girls, and where they increased, they did so at a faster rate than girls. This situation was reflected exactly in the changes in performance in algebraic problem-solving. The general conclusion drawn is that 'even though girls often seem to start ahead of boys in both the levels of their attitudes to themselves and mathematics, and indeed their actual performance level, they are eventually overtaken by the boys' (*ibid.,* p. 157).

Kaur (1990) carried out a study comparing the performance of girls and boys in Singapore on the UK-based Ordinary Level mathematics exam where she found that 'on the whole, the girls were out-performed by the boys' (p. 98). There was an element of choice in the exam and she notes that: 'It may be speculated from the preference of questions where there was a choice and also the topics on which one sex performed markedly better than the other, that Algebra and Graphs was the topic for the girls and Mensuration the topic for the boys' (*ibid.,* p. 110). She links this success with a perception of algebra as centring 'mainly on computational skills (substitution of numbers into an algebraic expression) and application of techniques such as factorization, drawing a graph, etc. on which girls appeared to be better in this present study' (*ibid.*).

SYMBOLIZATION

Mason (1987) argues that the attention of pupils must be drawn more explicitly to the power of symbols in order fully to understand why they are used and to become more articulate in using them and,

> in particular, the processes of specializing in order to find out what an abstract statement says in particular cases, and of generalizing in order to seek the generality being illustrated by particular examples. They can be shown by example that asking yourself what is common among several acts or situations can lead to generality. They can be shown by example that frequently stopping and mentally reconstructing what a text or class has been about helps them construct a reasonably coherent account of a topic, and that by trying to verbalize with the support of diagrams and metaphors they make their ideas accessible on a later occasion. In other words, they actually become aware of learning. (p. 213)

The apparent simplicity of the process being advocated here may disguise the difficulty usually associated with abstraction. However, the point Mason makes is a powerful one and is useful to remember when we consider the problems related to symbolization generally.

This point is taken up by Kaput (1987) who considers the meaningfulness of symbolic expressions as algebraic objects. He describes a symbolic expression or scheme as having, in itself, no meaning and requires a field of reference to give it a meaning. He refers to symbols as phenomenal objects (Mason uses the word 'images' to mean more or less the same thing) and sees these as 'stabilising the child's world' (ibid., p. 163). He notes that:

> the fact that mathematical abstractions, especially those that have the character of 'objects', such as sets and numbers, are *experienced* as real entities separate from oneself does not prevent us from regarding them as individual, personal constructions. In fact we see the heart of mathematical instruction to be in providing opportunities, mainly actions and reflections on those actions, for building phenomenological mathematical entities that, as Mason points out, become 'palpable' as they are used. Ultimately, for individuals they are the entities to which shared symbols refer. (*ibid.*, p. 175, author's italics)

As noted previously (Kieran 1989, 1992, Sfard 1991) it is the treatment of algebraic expressions as objects and according to structural rules that is a main difficulty in making algebra meaningful to children. The idea advocating that they stop and 'mentally reconstruct' the focus of action is highly important in order for any meaningfulness to be associated with the shared symbols.

Abstraction from data

Redden (1994) gave four number pattern stimulus elements to a sample of 1,435 children aged 10 to 13 years and asked four questions about each:

1. What is the next term in the pattern?
2. Describe a general rule for the pattern in natural language.
3. Calculate the value of an uncountable term (e.g. $n = 80$).
4. Write their rule in the symbolic notation of mathematics. (Redden 1994, p. 89)

Using the SOLO taxonomy (Biggs and Collis 1982) to evaluate outcomes, they identified two hierarchies of growth. The first was the *data processing dimension* where children exhibited an ability to handle a differing number of data elements (number facts) provided by the stimulus item. Within this hierarchy there were two levels:

(a) the concrete symbolic mode where children were able to respond to one piece of data within a stimulus element but could not give accurate descriptions when more than one bit of data was involved;
(b) a higher level, where they were able to process more than one piece of data successfully.

The second hierarchy related to the dimension of *expressing generality* where four hierarchical groups were identified:

(a) the pre-structural group where children did not engage successfully in the stimulus items (their work was not included in the overall analysis);
(b) the group who responded to the appropriate part of the stimulus item and gave a specific example but not a general description of pattern;
(c) the group that tended to focus on outcomes and to give a description of how to move from one term to the next in a sequence;
(d) the group who were able to provide an overview of data 'by describing a functional relationship between a dependent variable and an independent variable' (*ibid.*, p. 93).

These hierarchies provide a map for the progression from a procedural to a more structural perspective in the development of algebraic thinking in children.

Symbolizing a process

Reggiani (1994) studied the levels of generalization of 11- to 12-year-olds in Italy focusing on a problem given to pupils, the solution to which did not depend on data introduced. The problem was the following:

> Consider the following game: Think of a number, double it, add 5, subtract the number you thought of, add 2, now subtract the number you thought of again and then multiply the total by 3.
>
> Your friend says that whatever number you think of, the results will always be 21. Try and see if he is right.
>
> a) Start with number 7 and do the necessary calculations. Write the calculations.
>
> b) Now start with number 20 and then 100 and do the same thing. Write the calculations.
>
> c) Try to understand why you have always got 21 and write your observations (you can use schemes, letters or other symbols to help you.
>
> d) Was your friend right? Why? (p. 100)

She found that the children who had experienced no formal algebra were able to 'think algebraically' and 'sense the general' starting from particular cases (*ibid.,* p. 104). Some were able to explain their solutions verbally and symbolically, although at a naive level as, for example, the pupil who wrote 'I always obtain the same number because I multiply, add, subtract and add again to different numbers . . . always the same numbers, the results do not change' (*ibid.,* p. 101). There seemed to be an intuitive appreciation of the notion of generality, but the level of generalization reached was not considered stable. In the course of their calculations, Reggiani notes that at least 40% of her sample of 105 first-year secondary pupils used the equals sign in a procedural way (Sfard 1991).

Another study focusing on a different sort of stimulus for symbolization involved children in solving algebraic problems presented as short stories (Brito-Lima and Da Rocha Falcão 1997). Seventy-two children from 6 to 13 years old were interviewed and the resultant protocols were analysed. The researchers note that the youngest children produced what they call 'pre-algebraic schemes' which included 'hybrid equations, where natural language, icons, numbers and mathematical formal operators co-exist' (*ibid.,* p. 206). These performed a representational role and acted as guides in solving the problem which was seen as a precursor to the intermediate level where children were encouraged to model the situation however they liked, and then to write down the corresponding mathematical expression.

What is being symbolized?

An example of how the symbolization of a mathematical procedure overtakes the meaning of the focus of a mathematical problem is given by Underhill (1991) and illustrates Mason's point about the helpfulness of stopping to think and reconstruct when engaged in mathematical problem-solving. Underhill notes that when asked what is meant by 'average', pupils will reply in terms of the verbalized or symbolized procedure involved. In the case of the mean it would be $x + y + z/3$ where the three numbers to be averaged are x, y and z. The focus is the concept of average and he quotes an example of how an adult found it difficult to explain what was meant by the word 'average'. He first of all described it as the 'middle number'. He was then given a problem involving temperatures over a week and asked to find the average temperature and he correctly computed the arithmetic mean. He was then asked to add or subtract a few degrees, one day at a time, to see what effect that had on the average. Having done this, he arrived at the conclusion that 'If each number in a group is changed in the same way, then the representation of the group must also change' (*ibid.*, p. 245). Following another exercise, Underhill says 'He explained that his experience in averaging grades has taught him that because all the numbers in a group make equal contributions to the average, one number which is considerably higher or lower than the others can drastically affect the mean' (*ibid.*).

This is another example of the recapitulation called for by Mason (*op. cit.*). It is also an important reminder of the fact that symbolization is a means to an end. We have seen evidence of children's difficulties in learning algebra as the symbols become more removed from familiar contexts and become free-standing equations. Formulae, from their conception, are intended to represent situations but the meaning of these situations receded with mechanical use in problem-solving situations where their origins are easily forgotten.

A different sort of question related to what is being symbolized was explored by Furinghetti and Paola (1994) who used a questionnaire to probe the perceptions of students of 12 to 17 years of age relating to parameters and their relation to unknowns and variables. The result indicated that 69% recognized that there was a difference, but only 20 out of 199 were able to use a sentence to explain what they considered the differences to be. They also did not use basic notions such as generalization or the presence of quantifiers in their explanations. One student recognized that the difference between the three had to be stated by the problems.

Stereotypes of literal symbols

An argument is put forward by Bills (1997) related to the possibility of an unnecessary difficulty enforced upon pupils learning algebra by stereotyping the use of letters to the point where certain letters are always expected to fulfil certain roles. An example is given where, when a pupil is asked to factorize $x^3 - a^3$ he states that he does not known what a is. Up to that point, all polynomials that had been factorized were in x and had numerical coefficients. The immediate response here was that he wanted to evaluate the letter a but had no problem with the x. 'The role of x as a variable, that is as a quantity which can take any value and takes no particular value, was well established' (*ibid.*, p. 76). However, by using the letter a with the letter x as opposed to using either *b* or *y* was seen to relocate the task in a different context which confused the pupil.

In another example, pupils were asked to identify the equation of the tangent touching a circle with centre *(2,4)* at a point *(p, q)* lying on the tangent. All but one pupil gave their answer in terms of *x* and *y*. Bills makes the distinction between mathematically necessary and culturally determined choices and points out that the letters *x* and *y* carry with them many contexts and messages which may at times be useful to the pupil but also can prove obstructive. A similar problem was studied by Wong (1997). She set out to investigate secondary pupils' difficulties with exponential expressions containing numbers for indices compared with expressions containing letters for indices. She found that they performed better when the bases were letters and the indices were numbers than vice versa. For example, they could not translate the rule of adding the indices which were given as letters, when two exponential numbers were multiplied.

Symbolizing process and concept

Crowley *et al.* (1994) discuss the meaning of symbols within algebra and refer to the notion of a procept (Gray and Tall 1993, Gray and Tall 1994) which is 'a symbol that stands for both process and the output of that process' (Crowley *et al.* 1994, p. 241). It is argued that mental manipulation of procepts brings with it greater thinking power and flexibility which grows over time; as a result, gradually the process is itself compressed by the thinker to the extent where it becomes an arithmetic or algebraic object. The most obvious example is counting which is a process that eventually becomes compressed into the concept of number with number symbols. In this particular study they explored the notion of an algebraic expression as a procept:

> From our viewpoint ... children see *x* + 3 as a process and not a mental object – a process they cannot carry out because they do not know what *x* is. To be able to cope with such a symbol requires not only that it be given meaning, but the meaning should cope with it both as a process (of evaluation when *x* is known) and also as an object which can be manipulated as it stands. (Crowley *et al.* 1994, p. 242)

They had found in previous research that children were prone to *conjoin* terms, i.e. where they have 3*a* + 4*b* + 2*a* they will call it 9*ab*. For this reason, their sample in this study involved both a group of 13- to 14-year-old pupils and second-year student teachers, having considered that the latter group would be less likely to conjoin. They were given three addition and three multiplication questions which increased in difficulty, where the earlier ones suggested a syntax order but the more difficult ones did not. The results show that as the difficulty of the questions increased, both pupils and student teachers reverted to more process-oriented statements in which the operation appeared on the left and the answer on the right. Where the syntax was too complicated to allow for direct translation and did not encourage the use of letters as algebraic objects, errors occurred which reversed the role of the letters. The results show that the pupils did particularly badly when the syntax of the equations was altered, as, alas, did many of the student teachers (although it must be borne in mind that they had not studied mathematics for three years). The authors note that 'We see children responding to the translation process not only in purely syntactic word order, but also by attempting to make sense of the data using arithmetic and algebraic constructs related to their stage of development in algebraic sophistication' (*ibid.*, p. 246). By regressing to a procedural mode of operating algebraically, they make more errors.

Stacey and MacGregor (1994) investigated children's tendency to have a problem with using the notation for the algebraic product, i.e. *ab*, to denote the sum, $a + b$ (this is called a 'parsing obstacle' by Tall and Thomas, 1991 – see below). They concluded that many of the pupils' errors were due to 'drawing parallels with notation systems other than algebra', for example, personal shorthand system, and leaving out what may be unimportant words in another context (such as note-writing) (*ibid.*, p. 296). They note that 'students seem ... to often use a personal shorthand that only looks on the surface like algebra' (*ibid.*). They also conclude that rather than using a formal rule, pupils manage to achieve closure as a result of Gestalt.

Graphing

Chazan and Bethell (1994) have studied high school students' approach to drawing graphs and the often unconventional products which result. They suggest that such graphs 'challenge us, as teachers, to articulate and discuss with our students how graphs work in mathematics and the reasons behind the conventional approaches to creating "pictures" of the relationships between quantities in situations' (*ibid.*, p. 176). Their particular concern was to teach students across a wide spectrum of ability and 'to create a course which could be useful and challenging' for pupils who would likely finish their education at the end of high school (*idem*). They decided to do this by focusing on quantities that can be measured and counted and to investigate the relationships between them. Emphasis was placed on the fact that though graphs in themselves are not pictures, they 'picture' relationships. Pupils were asked to write a story, swap it with a partner and each draw a graph to illustrate the other's story. In the early stages it was found that the selection of a quantity in itself was quite complex; for example, in a story about a ship grounding, some pupils listed the time that various events happened as quantities, which did not fall into the researchers' preconceptions of a quantity that can be counted, and yet they had to acknowledge that this is one of the conceptions of quantity that pupils may bring to the situation. Some of the results included graphs with time on one axis and units on the other, and in some cases there was no indication of units at all. In one graph representing books being checked out at a library over an afternoon, some pupils had difficulty recognizing whether the decreasing parts of the graph represented the number of books being taken out or the rate at which they were being taken out.

The importance of a 'good picture' is emphasized by Chazan and Bethell.

> A good picture captures information about the rate of change of quantity as well as about its value. This emphasis on the graph as a picture of a kind, one which conveys information about the dependent quantity and how that changes, allows students, and us, their teachers, to argue for and against particular ways of making a graph. In particular, thinking about the types of pictures produced by mathematical approaches to graphing suggests reasons for having an equally spaced metric on the axes, reasons for accumulating the independent variable in a situation, and for choosing one quantity to be an independent variable and another to be dependent. (1994, pp. 179–80)

In other words, graphing activity is rich in potential as a teaching medium for a variety of aspects of algebra, and when pupils produce their own situation to represent and then to discuss the outcome, they are involved in reflecting upon their work and giving reasons for what they have done. In the course of this, they are learning important mathematical concepts in a meaningful context. The conclusion of the researchers was that they did, in fact, want their pupils to think of graphs as pictures 'but as pictures which must be drawn in certain conventional ways' (*ibid.*, p. 183). More importantly perhaps was their discovery that pupils were often drawing 'correct' graphs 'for a different relationship between quantities than the one they were being asked about' which emphasizes strongly the crucial need for pupil–teacher discussion and also the crucial nature of it (*idem*).

The relevance of context in graphing gains further support from Ainley (1995) who found that 'When presented with purposeful tasks in a familiar context the children seem to be able to act intuitively to use line graphs' (p. 16). This observation arises from her work with the Primary Laptop Project in the UK. She suggests that the difficulties children have had with respect to graphing reported in research in the past may have been due to pedagogical causes rather than cognitive difficulties on the part of the children.

THE USE OF TECHNOLOGY

Algebra is an area of mathematics that attracts a great deal of attention where the use of technology in the teaching and learning of mathematics is concerned, and where it is seen to have great potential (Monaghan 1995). While there are many studies that appear to indicate there are advantages in using technology in the teaching of algebra, it may be salutary to note Pimm's cautions in this respect. Pimm (1995) argues that 'Ironically, technology is being used to insist on screen (graphical) interpretation of algebraic forms. There is a strong presumption that symbolic forms are to be interpreted graphically, rather than dealt with directly' (p. 104). He points to the promotion of the concept of function in particular through the medium of technology and suggests that the development of this perspective is contrary to the historical development of the subject (see Kieran 1992). He believes that the ultimate effect is to 'switch from description to prescription: from the fact that it is possible to describe one thing in terms of another, to the fact that it must so be seen' (*ibid.*, p. 105). Certainly the number of studies using computer technology that focus on the concept of function are numerous (e.g. Crawford *et al.* 1994, Confrey 1994, Yerushalmy 1997, Gomes-Ferreira 1998). Pimm seems to be suggesting that the view of algebra that is conveyed in the context of technology becomes rigid and static and loses what for him is its essentially transformative power. Some of the studies reported below indicate where graphic calculators and computers have had some positive outcomes, but there are others like Pimm who consider that algebra must be engaged in more directly, at least in the earlier stages of learning.

Yerushalmy (1997) takes a different stance and views technology as a medium for bringing the real world into the classroom and offering a variety of options for solving word problems as well as a means of modelling problem situations. He sees the role of technology not simply as allowing a change in the sequence of learning but 'a deep change in the scope of traditional actions and objects' (*ibid.*, p. 167). He suggests that 'viewing algebra as a study of functions aided by technology, shakes common models of problem solvers, particularly those of weak students.'

Teaching functions

Confrey (1994) explored the use of multi-representational software to teach the transformation of functions at a variety of levels. The software involved the use of four different representations: graphs, tables, algebra and calculator keystroke notation, and the aim was to focus on the meanings of transformations in the different settings. The system allowed a record of the pupil's actions to be made. Each of six approaches used was at a different level of sophistication and ranged from substitution to a template (which essentially allowed pupils to identify the actions linked with the parameters), equations such as $y = Af(Bx + C) + D$ to scaling the axes when graphs were used. The approaches described were developed over a period of extensive experimentation with pupils, and Confrey concludes that they all have advantages and disadvantages and recommends that they should be used according to which is best for a 'particular functional family, context or student preference' (*ibid.*, p. 224). Analysis of the data from the detailed recording of pupils' actions throughout provided many insights into their thinking in an algebraic context, thus allowing a deeper consideration of 'ways in which learners structure their own learning as well as of the ways in which the setting structures it' (Noss and Hoyles 1996, p. 108). While this work was carried out mainly with college students, it provides an example of how a technological environment can contribute both to extending algebraic knowledge as well as providing data for obtaining deeper insights into the learners' algebraic thinking.

In another project using computers in the teaching of functions (Kieran 1994), the view of function is not only as one where the letter is viewed as a variable but 'with representations that do not involve symbols, such as story contexts' so that 'the functional concept is extended to include the expression of variable notions in a more rhetorical formulation' (*ibid.*, p. 162). The theoretical underpinning of the project emphasizes the process–object duality within mathematical conceptions and it was decided to have two versions of the projects, one using software that was a process-oriented version beginning with an operational approach, and the second an object-oriented version beginning with a structural approach. She states that 'More recent research that has taken advantage of the potential of technology has pointed to ways in which we might rethink the teaching of algebra, but much of that work has lacked a theoretical framework by which we can not only assess what is occurring in the novel environments but also to place it in a broader context' (*ibid.*, p. 173).

Promoting versatile thinking

Tall and Thomas (1991) offer the hypothesis that 'as soon as children are unable to give meaning to concepts, they hide their difficulties by resorting to routine activities to obtain correct answers and gain approval' (p. 127). They explored this hypothesis in the course of a study in which pupils used the Dynamic Algebra Module with the aim of promoting more versatile thinking in the context of learning algebra. They identified four kinds of obstacles frequently met by pupils:

1. the parsing obstacle – e.g. *ab* means the same as $a + b$;
2. expected answer obstacle;
3. the process-product obstacle, e.g. $2 + 3b$ represents both the process and the product;
4. the lack of closure obstacle – discomfort in handling an algebraic expression that they cannot 'carry out' or for which they cannot reach what is, for them, a recognizable conclusion.

They worked with 11- and 12-year-olds in a middle school and followed them on to the first year of secondary school. Each pupil in the sample was paired with one who was not using the DAM. The module sets out initially to tackle the problem presented by a letter being used as a label for a number or set of numbers, and then establishing the equivalence of expressions through evaluation. The results showed that the children taught with the DAM:

1. were able to cope with the parsing and product-process obstacles;
2. were more likely to be able to think of expressions as obstacles and hence overcome the problems presented by lack of closure;
3. appeared to have no trouble with the expected-answer obstacle.

Tall and Thomas conclude that children using the DAM showed more versatility in their thinking than pupils following the more traditional course. The computer environment appeared to promote this versatility and allowed the pupils to deviate from routine actions and become more explorative in reaching their solutions.

Graphic calculators: the variable and algebraic language

The graphic calculator was used to introduce the concept of variable to children aged 12 to 14 years in a study undertaken in five schools in the United Kingdom by Graham and Thomas (1997). Previous studies (Thomas and Tall 1988, Tall and Thomas 1991) had led them to conclude that 'it was possible to improve students' understanding of variable by giving them environments in which they could manipulate examples, predict and test and gain experiences on which higher-level abstractions could be built' (Graham and Thomas 1997, p. 10). The value of using a graphic calculator was seen to be that it intrinsically uses variables in its operations and that the multi-line display allows the user to see, reflect upon and react with several sequential inputs/outputs. Five teachers chose one class to teach the selected algebra module and a second class was used as a control group. Both sets of classes were given pre- and post-tests. The results showed a relative improvement in the performance of the classes using the graphic calculators, with post-test scores being significantly better than control classes in four of the five schools. Four of the five questions where they performed significantly better required an understanding of the use of a letter as a generalized number. The conclusion was that 'Approaching algebra by gaining an appreciation of the use of letters as labelled stores will, we believe, help students construct an understanding which will improve assimilation of later concepts' (*ibid.*, p. 15).

Cedilli Avalos (1997) reports a study with 11- and 12-year-old Mexican children which involved the use of the graphic calculator to introduce and teach algebra as a language. The aim was to investigate the extent to which the calculator 'language' viewed as a means of expressing rules governing number patterns, could help children to grasp the fact that the algebraic code can be used as a tool to help in the solution of problems. Evidence was gathered that suggested that children's performance on given algebraic tasks using the graphic calculator helped them to gain an awareness of the generality of algebraic expressions and to proceed to some understanding of inverse functions and algebraic equivalence.

Logo and algebraic language

The benefits of a computer environment for helping children to deal with the language of algebra were also noted by Harries and Sutherland (1995). They used Logo in a study and argue that it allows greater emphasis on the language and structure of algebra and less on the output which, if this is the case, could allow for the more meaningful understanding of structural aspects of algebra that appear not to be addressed in much of algebra teaching (Kieran 1992). In the Logo program used, the emphasis was on the formulation and manipulation of algebraic objects but not on immediate solutions. The study involved a group of 13- to 14-year-old pupils in a comprehensive school in the United Kingdom who, in the normal course of events, would not have been expected to progress much beyond arithmetical problem-solving. The program was chosen because it was seen to provide pupils with access to algebraic processes such as generalizing and transforming algebraic symbolic objects, and to a system of symbolic representation which involved the use of 'algebraic sentences'. The main aims of the study were to investigate:

1. the algebraic understanding of the group as they pursued the Logo course;
2. their use of Logo in non-Logo contexts;
3. the relationship between the development of algebraic understanding and knowledge of arithmetic (focusing on informal methods).

The example of the case study given indicates that the difficulties encountered included problems with the concept of equivalence and confusion with working between objects and processes. The results have helped to focus on specific concepts such as the variable which needed to be developed further within the context of the Logo program. Once again, the main difficulties that have been isolated are identical to those that occur elsewhere in the literature.

IMPLICATIONS FOR TEACHING

While it is not feasible or desirable to draw conclusions from a brief review of some of the body of research related to algebra in the school curriculum, it is clear that there are some points that recur throughout the literature. Some of those which may have particular importance for the teaching and learning of algebra are given below and are grouped under four broad headings.

Pre-algebra

1. The teaching of algebra begins at primary school level when children are taught basic concepts such as equivalence and the notion of an unknown, and are introduced to the symbols for them. The structural rules that relate to the use of the equals sign need to be recognized and used so that misunderstanding and misuse can be avoided in proceeding from the arithmetic to the algebraic level. In particular, children need to recognize that the equals sign does not necessarily indicate that a numerical answer is required. They also need to be reminded of the requirement of balance on either side of the equals sign in any expression containing one.
2. There appears to be a need for more attention to be focused on the different uses of letters or other symbols to represent unknowns in the earlier years; in particular the recognition that it can have a fixed value (as

in $2 + \Box = 5$) or it can be a variable (as in $2l + 2b$ when describing how to find the perimeter of a quadrilateral).

3. A greater awareness of the difference between the *procedural* and the *structural* aspects of the manipulation of numbers as literal terms in the progression from arithmetic to algebra could help promote the understanding of algebraic expressions as *entities* or *objects* in their own right. This appears to be the most crucial, but also the most difficult, aspect of children's success in learning algebra. The importance of the *order of operations* needs to be established more firmly in order for this to happen.

4. The alternation of the *arithmetic-to-algebra* and the *algebra-to-arithmetic* pathways appears to be particularly important throughout this phase of learning algebra. This helps to remind children of the connections between the two, and the changes that are involved from one to another.

Solving equations

1. For the notion of an equation to be meaningful to children, they need experience of *generating their own equations* from situations with which they are familiar. The evidence suggests that learning to solve problems by mechanical substitution does not lead to meaningful learning.

2. A greater awareness of the ways in which children use *procedural or operational (as opposed to structural) methods* to solve equations is needed to help them to recognize that these methods will let them down as they move on to deal with more complex equations. It would appear that most children do not progress beyond using procedural methods, but to accommodate those who may, these limitations need to be made explicit.

3. Again, if the *order of operations* has not been firmly established at an arithmetic level, children will have difficulty in manipulating algebraic expressions in the course of solving equations. To achieve this, the evidence suggests that children need to learn to concentrate on the *whole of an algebraic expression* and not follow a left-to-right approach in reaching a solution.

Solving word problems

1. The importance of *language* is highlighted throughout the research on solving word problems, both in the way in which it is used to present problems and the part it plays when children try to verbalize given problems in their own language. If they have not generated a problem themselves, then they have to use their own words to translate the problem into *a form which is accessible* to them.

2. The establishment of the concept of the *part-whole relationship* appears to have positive outcomes when helping children to learn to solve word problems. This helps them to identify the relationships between the constituent parts of a problem.

3. There is a tendency for children *to use all the data* given in a word problem

whether or not it is relevant to solving the problem. This is where using a strategy such as the part-whole method can be useful in helping them to analyse a problem.

4. *Modelling* a word problem helps children to understand symbolic representation by involving them directly in the process of symbolization so that they are able to relate its meaning to their own experience.

Symbolization

1. There is evidence of considerable rigidity in children's *interpretation of symbols* when they adopt stereotypical views. This applies to literal symbols such as *x* and *y* and to '=' which, from a procedural perspective, they take to mean that a single result is expected.
2. The research suggests that for structural understanding of algebra, children have to come to see *algebraic expressions as entities or objects*. For this to happen, they must focus on the whole of the expression which is to be 'used' as an entity. This may cause confusion with the opposite situation that prevails in solving word problems and emphasis is laid on the part-whole approach and a single answer is expected.
3. Children have a stereotypical view of the way in which they expect an equation to be represented and will carry out unnecessary processes to try to get an equation into that format.

The use of technology

1. There are arguments for and against the use of technology in teaching algebra. However, there is evidence that it can be effective in the way in which it helps children to *visualize various processes* and gain a *more active perspective of the subject*. This applies to aspects such as modelling word problems and generating graphs of families of functions.
2. Much of the research using technology in the teaching and learning of algebra has focused on *functions,* some of which has led to the suggestion that the use of technology in teaching the concept in this way may help learners to understand the *structural aspects* of algebra more easily.

Chapter 5

Classroom Research

INTRODUCTION

At the beginning of this book we noted the changes in perspectives on research in mathematics education. The considerations that led to these changes are also those that have led to an increased focus on classroom research. While learning is seen to be the construction of the individual child, the emphasis is now placed on the situatedness of the learning (Lave 1993, 1996, Lave and Wenger 1991) and the part that peers, teachers, the classroom setting and society at large have to play in that learning (e.g. Mellin-Olsen 1987, Nickson and Lerman 1992, Rogers 1992, Noddings 1993, Skovsmose 1994). In this chapter, we shall consider research that relates to the mathematics classroom as a context for individual learning in which the pupil and the teacher are seen as the focal points against the background of the class as a whole. First, however, we shall look briefly at the idea of the culture of the mathematics classroom as a phenomenon on its own and consider the theory that informs this focus.

CULTURE OF THE MATHEMATICS CLASSROOM

Culture is a concept that now pervades mathematics education at all levels (Bishop 1988, Nickson 1992). At the broader level, it has been considered with respect to ethnicity (D'Ambrosio 1985), politics (Fasheh 1982, Joseph 1990, Mellin-Olsen 1987, Skovsmose 1994) technology (Noss 1988), cross-cultural issues (Zaslavsky 1989), mathematics curriculum development (Nielsen, Patronis and Skovsmose 1999) and multicultural mathematics (Nickson 1988). Feiman-Nemser and Floden (1986) note that when we talk about culture, this 'implies inferences about knowledge, values, and norms for action, none of which can be directly observed' (p. 506). Referring to cultures of teaching they suggest that this is a new way of examining educational practice and it is one that has taken hold in mathematics education in particular in recent years, to which the studies in this chapter bear witness. By the culture of the mathematics classroom we mean the invisible and shared meanings that are governed by norms and values that children and teachers bring to the classroom and, in turn, govern their actions and interactions in it (Nickson 1992). It is probably fair to say that twenty years ago, virtually none of these terms such as culture and shared meanings would have been associated with mathematics education in particular, but the fact that they now do indicates, first, that mathematics education has emerged as a discipline in its own right, and second, that mathematics educators have become more aware of the responsibility of their task in teaching their subject effectively to young people who must make their way in a technological world.

It is clear from the description of culture just given that there are as many cultures of the mathematics classroom as there are such classrooms. The composite of knowledge, beliefs and values that any one class will have is a unique entity, but what these classrooms all have in common is the subject of mathematics, and it is mathematics that will exert a common influence. A concern with the culture of the mathematics classroom is, then, to ask questions about the effect the subject has with respect to the role of the learner, the teacher and the social unit of the class. It is easy to assume that mathematics is in this context a unifying agent, but this is not necessarily the case because of the variation in perspectives both of the subject and of the roles each of the players in the classroom has of what mathematics teaching and learning are all about (*ibid.*). The teachers who view mathematical knowledge as an inert body of received wisdom will approach their teaching in one way, while the teachers who view it in more open terms both epistemologically and as a social tool, and who see it as necessary for the personal autonomy of all pupils, will approach it in another (Lerman 1990, Nickson 1992).

The studies considered in this chapter are grouped under two main headings: pupils and teachers. The third obvious focus here would be the classroom, but the nature of the current research pattern makes this level of categorization almost impossible to achieve. However, the perspectives of the teacher and the pupils are more manageable and help to make sense of this body of research as a whole. The first perspective to be considered is that of the pupil.

THEORETICAL CONSIDERATIONS UNDERLYING CLASSROOM RESEARCH

The theoretical shifts already noted in Chapter 1 indicate how perspectives of teaching and learning mathematics have altered in recent decades. While there have been changes in the classroom that reflect these recent theoretical shifts to some extent, the changes within the research methodologies have been more marked. The most obvious change has been the increase in the use of ethnographic approaches (Eisenhart 1988) which are qualitative in nature and use an interpretive paradigm which is more appropriate to the investigation of the social context of the mathematics classroom, as well as factors such as the values and beliefs that teachers and pupils bring to it. Ethnographic research has been described as being concerned with learning *from* people rather than the study *of* them (Spradley 1980) and the methodologies usually used for data collection within this paradigm include participant observation, case studies, questionnaires, interviews and collecting artefacts (written documents and other forms of 'hard copy') related to the topic of study and researcher introspection, all with a view to providing 'a comprehensive picture of the complex meanings and social activity' that is being studied (*ibid.,* p. 107).

An overview of the growth of research in mathematics education helps to see how this gradual swing from a quantitative to a more qualitative methodology has evolved. In a résumé of different levels at which research in mathematics education has been undertaken over the years, Koehler and Grouws (1992) noted four levels of development. Early research at Level 1 focused on teacher effectiveness where either a particular component of teaching or the individual teacher was studied in order to isolate factors that either promoted or hindered effective teaching and learning. At Level 2, there was a move towards multiple classroom observation and 'process-product' research which focuses on classroom processes in an attempt to identify what goes on in a mathematics classroom. This kind of research tends to be characterized by the frequency of particular behaviours on the part of the children or the teacher within a mathematics lesson. The next stage of development broadened the parameters yet again when, at Level 3, it was recognized that 'Pupil characteristics such as gender, race, and confidence level can affect teacher practice and pupil actions' (*ibid.,* p. 116). Although a 'process-product' perspective still holds, factors such as pupils' attitudes are taken into account. The highest level of complexity is Level 4 where new factors such as the wider range of mathematics being taught to a wider ability range and an increased variety of methodologies have to be considered. The greatest change at this level is that research focuses on teaching and learning at the same time (*ibid.*). Developments in cognitive psychology have played a major role at this level of research (Putnam, Lampert and Peterson 1990) and it might be defined as 'research that has a strong theoretical foundation and is based on a model that involves many factors' (Koehler and Grouws, p. 117).

A large part of the research in mathematics education currently takes place at Level 4. From the pupil's point of view, the influence of teacher actions is no longer the main focus, but other matters such as pupil attitudes and beliefs in relation to themselves, the teacher and the subject are all of concern. Apart from what they bring to the classroom situation as unique individuals, pupils may find their expectations challenged with respect to what mathematics teaching and learning is all about. If the messages they receive are that they are expected to participate actively by contributing their own ideas, and even that they may challenge those of other pupils as well as the teacher, this may well produce a kind of dissonance that could obstruct their learning if they have a more traditional view of what teaching and learning is all about. Where the teacher is concerned, issues have broadened to include matters such as their beliefs with respect to the nature of mathematics, their attitudes, and their beliefs about their role as teachers. Koehler and Grouws also point out that: 'Teacher behaviour is influenced by the teacher's knowledge of (a) the mathematics content being taught, (b) how students might learn or understand that particular content, and (c) methods of teaching that particular content' (p. 118).

It has been suggested that the potential for research probing this area of teacher education is considerable and can provide a deeper understanding of teachers' meaning-making processes to form a basis for teacher education that is responsive to their beliefs and needs (Brown *et al.* 1990). The holistic approach adopted at Level 4, then, has potential for wide-ranging effects for all of the participants in the mathematics classroom, for methodologies and for the way mathematics itself is perceived, taught and learned. Whatever approach is adopted in undertaking classroom research, the theoretical approaches discussed earlier in this book clearly will inform the methodology selected. Adopting the Cognitive Guided Instruction perspective, for example, will mean that research informed from this perspective might look at how children do mathematics in a classroom setting quite specifically to identify the strategies used in their learning. A constructivist would be likely to focus more on the various levels of discourse within a classroom and on the consensual mathematical meaning-making of the participants as exemplified by Goodchild (1995) in his study of a Year 10 mathematics classroom in the UK. Someone engaged in research related to realistic mathematics education would also focus on how children engage in mathematics as a collaborative activity, but perhaps might be concerned more with the form that the children's understanding of the relevant mathematics is taking. There are, however, some emergent theories that are over-arching and more global, that help to inform the analysis of what is taking place in a classroom in a more general way. One such example is the didactic contract.

The didactic contract

An example of how research paradigms may shift in the course of the development of a study is provided by Schubauer-Leoni and Perret-Clermont (1997). They have studied social interactions and mathematics learning over a period of years during which time they note that 'The visible outcome was sometimes answers to the questions raised, but often also of the *modification of the research questions,* and as a consequence the search for changes in *methods appropriate* for their investigation' (*ibid.,* p. 265, authors' italics). They moved from a stage of asking questions related to the benefits of peer interaction in children's learning of elementary mathematics to a stage where they became interested in 'the *experimental history of the* subject' which allowed a more dynamic approach to be adopted in their research (p. 272, authors' italics). The stage their research has currently reached is described in terms of the paradigm of the *didactic contract.* This is a concept developed by

Brousseau (1988) and Chevallard (1988) in which the 'triangular nature of the teaching relationship . . . is bound by an implicit "didactic contract" – that is a system of reciprocal and specific expectations with regard to the knowledge taught' (Schubauer-Leoni and Perret-Clermont 1997, p. 272). They wanted to study the dynamics of the interaction of children with each other as well as with their teacher with the intention of formulating a detailed description of 'the *simultaneously interactive and representational* nature of the processes involved' (p, 272). The assumption is that each facet of the teaching–learning situation is represented in different ways by each of the participants in the situation. They used the idea of *experimental staging* to study the broader situation in which the individual pupils and the teachers enacted their roles in the teaching–learning classroom context.

The interlinked phases of a didactic contract drawn up from the research were as follows:

> *Phase 1*: the teacher conceives of and writes up a mathematical task for the pupils (typical task within the current didactic contract) and is invited by the researcher to explain the rationale for the choices that were made.
> *Phase 2*: the teacher makes predictions as to the behaviour and performance of each pupil with regard to the task.
> *Phase 3*: the teacher administers the task to the pupils as usual.
> *Phase 4*: the teacher corrects the answers as usual. (*ibid.,* p. 274)

They note that pupils' views are affected by the implicit knowledge they have learned in the usual functioning of the didactic contract in the classroom and by the habits they have developed in the course of this classroom experience.

The researchers were interested to study the didactic contract from the teacher's viewpoint as a result of their previous findings in which they had noted differential response on the part of teachers to pupils in their classes. To undertake such a study they formulated four further stages of the didactic contract which deliberately set out to *break* it. These were as follows:

> *Phase 5*: the researcher submits to the teacher a task that is unusual for, and at the outer boundaries of, the didactic contract in force. The teacher is asked to determine the interest and feasibility of the task for this class.
> *Phase 6*: the teacher formulates the expectations of each child's behaviour and performance with respect to this unusual task.
> *Phase 7*: the teacher administers the task to the pupils.
> *Phase 8*: the teacher corrects the answers to this second task. (*ibid.,* p. 274)

Their results led them to the following conclusion:

> Above all we observed that teachers differentiate between cognitive levels of responses to the task according to their representation of each

pupil; the didactic contract appears to be the result of differential expectations by the teacher who, without realizing it, creates expectations in mathematics that are part of the sociological characteristics of the pupils in the class. The teachers tend to overestimate the performances of pupils from privileged social backgrounds and underestimate those of the pupils from underprivileged ones. (p. 274)

They note that the differential approach adopted here was not intended by the teachers and that they viewed themselves as 'committed to the cause of equal opportunity and democritization of access to school and knowledge' (*idem*). Their research as a whole has emphasized the extreme complexity of the didactical situation and the misunderstandings that arise from the implicit references of the participants to their different realities, or from an assumption about common experiences throughout their teaching–learning experience.

Other studies (with 7- to 8-year-olds and 11- to 12-year-olds) focusing on the pupil's perspective of the didactic contract were carried out in which a pupil took on the role of the teacher. They were asked to think up a mathematical task and to choose another member of the class to whom they would 'teach' the task. Three points emerged. They found, first, that the tasks chosen emphasized calculation. Second, their questions were formulated on the basis of the answers they would expect to get, a factor which was interpreted as assuming what they saw as their 'usual' role of responding to questions in class. Finally, a weaker classmate was chosen to take the role of their 'pupil' which was seen as necessary for the legitimacy of their perceived role of the teachers and allowed them to distance themselves from the 'pupil' in an authoritative way.

From the extensive body of research with the notion of a didactic contract at their centre, Schubauer-Leoni and Perret-Clermont (1992) have attempted to identify more specifically what has emerged from these studies and how it can progress the study of the teaching and learning of mathematics in particular. They identify two possible new theoretical fields, the first of which is *theories of the subject* which acknowledges the contextual situation for the knowing subject 'that no longer place them in isolation but situate them in relationship with their social and cultural universe' (*ibid.*, p. 279). This field focuses on the individual (teacher or pupil) via the situation. The second is a *theory of situations* which is concerned with the identification of situations in which the intention is to 'bring about the emergence of expected knowledge' in the individual, and here the focus is on the situation via the teacher or pupil (*ibid.*, p. 279). Mathematics as a discipline comes into play because at the centre of this field is the aim of trying 'to understand which situations establish which relations with the object of knowledge' and is concerned with the inter-relationships between different approaches used, the cognitive outcomes and abilities involved, as well as the cognitive tools and the role of memory (*idem*). The researchers point out that these fields (theories of the subject and a theory of situations) are not mutually exclusive but that it is productive to work with the two interweaving with each other.

In identifying two new theories or fields of study as they have done, Schubauer-Leoni and Perret-Clermont explicitly draw attention to the two main foci for classroom research which characterize many current research studies. The idea of theories of the subject (i.e. the individual being focused upon at any given time) highlights the situatedness of learning which has been referred to in these chapters (see Cobb *et al.* 1991, Lave and Wenger 1991, Boaler 1993). A theory of situations brings mathematics specifically into the arena and the concern becomes directed towards the 'bringing about' of mathematical knowledge in different situations as they are defined by the factors that make up the context of the mathematics classroom, including how the subject is taught, the skills and knowledge brought into play, how they are used and what the outcomes are.

Didactic engineering

The notion of didactic engineering has evolved from the idea of a didactic contract and provides both a theoretical framework for the teaching and learning of mathematics as well as acting as a research method. Douady (1997) describes it in these terms:

> Didactical engineering is both a *product,* resulting from *a prori* analysis
> – and a *process,* resulting from an adaptation to the implementation of
> the product in the dynamic conditions of a classroom. (p. 373, author's
> emphasis)

The approach encompasses the idea that teaching mathematics is based on choices which are analysed, argued and justified by research findings from both the cognitive and classroom perspective, i.e. from the point of view of the learning that takes place or the context in which it happens. Thus it can provide a framework either for the planning of teaching–learning situations or for their analysis. The didactical contract thus is viewed as 'the implicit or explicit agreement between teachers and pupils about what it means to learn mathematics and what role this learning plays in their relationship – because the didactical contract can sometimes get in the way of learning' (*ibid.,* p. 374).

Mathematics can play different roles within the teaching–learning situation and they are described by Douady as follows:

1. *Knowledge is what is at stake for the teacher but not necessarily for the pupils.*
   In this situation, the consideration is with pupil expectations, what it means for them to go to school and for what proportion of them knowledge is not the focus in the mathematics classroom.
2. *Knowledge is not at stake for the teacher or the pupil.*
   This pertains in a situation where the content of mathematics lessons is a

mechanical exercise in reproducing algorithmic versions of mathematical content over which neither the teacher nor the pupil have any ownership. However, by adopting this view, 'For the teacher and also for the pupils, survival is ensured' (p. 377).

3. *Knowledge is at stake for some pupils but not the teacher.*
   This is the situation when the teacher holds the view identified in (2) but some of the pupils may be genuinely interested and actually want to learn some mathematics in a meaningful way. However, as a result of the teacher's perceptions and actions, they become disenchanted not only with mathematics but possibly with school as a whole.

4. *Knowledge is at stake for both teacher and pupils.*
   This is the ideal situation but does not necessarily imply that knowledge is built upon and extended. This is where relevant research can inform the teacher's actions to help bring about effective learning.

An example is given by Douady (1997) of an algebraic topic being taught using the didactic engineering approach which she describes as an example of 'the tool-object dialectic' (p. 400). This describes the interplay of the various means already known (the 'tools') by which the algebraic content (the 'object') is accessed by the pupils and the teacher and gradually reaches a stage of 'familiarization' where pupils can use it in new situations. In the course of the lessons she describes how processes of contextualization, changing contexts, reformulating problems, decontextualization all take place. Pupils and teacher are both involved with making different aspects of these procedures 'their own' by relating them to what they already know and accommodating them in such a way that they are extended. What was new becomes a 'tool' and a dialectic is set up with another 'object' in the form of a new aspect of mathematical or didactical knowledge. Thus throughout these procedures, the teacher is transferring 'tool status into object and vice versa' (p. 406).

A Vygotskian perspective

The work of Vygotsky has been invoked in a variety of research studies related to mathematics education in recent years (e.g. Bartolini-Bussi 1991, Lerman 1992, Boero *et al.* 1995). Mellin-Olsen (1987) was probably the first to relate Vygotsky's work specifically to mathematics education, particularly his notion of *activity theory* which Mellin-Olsen describes as embodying both society and the individual: 'the individual acts on her society at the same time as she becomes socialised [into] it' (*ibid.*, p. 33). It is the involvement of the social dimension in the child's education that has made Vygotsky's work of particular relevance to the study of the teaching and learning of mathematics. It is helpful to examine Vygotsky's (1962) interpretation of the notion of imitation in connection with education to understand how this social perspective arises in the first instance. He notes that 'To imitate, it is necessary to step from something one knows to something new. With assistance, every child can do more than he can by himself – though only set by the limits of his development.' He goes on to develop the importance of this interpretation in the education of children and, having noted the importance of both imitation and instruction in the child's development, he says:

> What the child can do in co-operation today he can do alone tomorrow. Therefore the only good kind of instruction is that which marches ahead of development and leads it; . . . instruction must be oriented towards the future, not the past. (Vygotsky 1962, p. 104)

He suggests that in more traditional methodologies when we have asked children to do a particular task that they are unable to do without help, we have 'failed to utilize the zone of proximal development and to lead the child to what he could not yet do. Instruction was oriented to the child's weakness rather than his strength, thus encouraging him to remain at the preschool stage of development' (*ibid.*, p. 104). The zone of proximal development (now referred to by many as zpd) is described in terms of the discrepancy between what a child is able to do at the point of entry into a problem situation and the level reached in 'solving problems with assistance' (*ibid.*, p. 103).

Lerman (1998) elaborates on this Vygotskyan perspective and relates it to current developments in mathematics education. He suggests that with respect to zpd, Vygotsky 'probably intended to emphasize jointly the roles of the learning activity and the learning potential of the child' but notes that many tend to view zpd as 'a kind of force field which the child carries around, whose dimensions must be interpreted by the teacher so that activities offered are within the child's range' (*ibid.*, p. 71). However, drawing on the work of Davydov (1988, 1990), Lerman suggests that this was not the intended meaning of the zpd, and that it is in fact 'a product of the task, the texts, the previous networks of experiences of the participants, the power relationships in the classroom' (p. 71). Thus the zpd belongs to the classroom, not to the individual child, and it becomes a useful tool for the researcher's analysis of the learning interactions that take place in the classroom since it incorporates, amongst other things, the strategies teachers use in bringing it about. Lerman uses the Vygotskyan stress on social considerations to develop the notion of a 'discursive psychology for mathematics education' which he implies could act as a counter-balance to constructivism. He argues that 'A psychology focused on the individual making his/her own sense of the world does not engage with social and cultural life: other theoretical discourses such as approaches to sociology which merely describe, are not adequate for mathematics education either' (*ibid.*, p. 77). Rather, he suggests, what is needed is a theory which 'incorporates actions, goals, affect, power and its lack, based on sociocultural origins', a theory such as that which he has proposed and which is informed by the social nature of Vygotskyan thought (*idem*).

PUPILS IN THE MATHEMATICS CLASSROOM

In much of the work that has already been considered in previous chapters, we have seen that children are viewed as constructors of knowledge, although within that view there is a variation of approaches as to how this is interpreted in the classroom. The idea of children as 'individual constructors' implies a somewhat solitary and isolated situation, although this is not what is intended in theory, and need not necessarily be the case. Most of the studies that follow focus on children working in pairs or groups and their talk is very often the focal point of the studies. The negotiation and shared meanings which are implicit in such talk have resulted in the stress on language, but it also has to do with the importance of the role of language in concept formation, as we shall see.

The primary years

Collaborative problem-solving

An example of how Vygotskyan thought informs the interpretation of teaching and learning in the mathematics classroom is elaborated by Zack (1994) in a study of how children use each other's explanations to reach mathematical understanding. The study is built upon the notion of the zone of proximal development as 'the difference between the learning an individual can construct on her own and what she can accomplish with the assistance of someone else (a teacher, a parent, or a knowledgeable peer)' (*ibid.*, p. 409). The class was taught mathematics in 'half groups' of twelve each and children worked collaboratively in a classroom, first with a partner and then in groups of four, to solve problems which ultimately were presented to the larger group of twelve. The focus on peer explanation arose from previous observations of the class when it was recognized that these were seen as helpful, and particularly when what was offered was a 'clear explanation' (*ibid.*, p. 410). The intention of the study was to investigate what constitutes a 'clear explanation' as the teacher-researcher recognized that what she saw as 'clear' did not always tally with what her class labelled 'clear'. To explore the issue, it was decided to ask children to identify:

(a) helpful explanations;
(b) the children who gave these helpful explanations; and
(c) what kinds of help the children valued.

Children were asked to supply a written response to the following:

> Was there an explanation given by a fellow student that helped you to understand the problem? _____ or helped you to understand the problem better? _____ If yes, tell me the student's name. If you can, tell me why or how the explanation has helped. (*ibid.*, p. 411)

Two ways of collecting this information were tried; in the first instance, children were asked to undertake this retrospectively (some weeks later) and in the second, they were asked do so immediately after a problem-solving session. The number of explanations offered were in the ratio of 4:1 in favour of the second approach, indicating the importance of the immediacy of collecting data of this kind.

Four categories of explanations emerged, although it is noted that these are not discrete, but are inter-connected and can overlap with each other. The categories were identified as follows:

1. *Parameters or conditions of the problem* – explanation helped to identify what the problem required and the conditions imposed by it.
2. *Factual information* – explanation provided information needed to solve the problem.
3. *How to solve the problem* – (a) how to identify strategies to be used and to organize information; (b) how to represent ideas with mathematical symbols (e.g. symbols for fractions or a right angle as a right angle in different positions); (c) focusing on essence of the problem (identifying what is central to it).
4. *Presenting an alternative solution to the problem* – a simpler or more economical solution is offered.

There were 'explanations' that were too vague to be categorized, such as 'They helped me to understand it better.'

In her discussion of these results, Zack (1994) notes that, although aware of being helped by an explanation, pupils would frequently identify only one component of the explanation. The results indicated that explanations from the children who were 'more adept' were nominated more often than those of others and she cites Sierpsinka (1992) in explaining this phenomenon: 'understanding is irrevocably linked with . . . explaining (which has traditionally been separated from understanding)' (p. 78). The possibility of many votes being allocated to the pupils with the right answer is acknowledged since the nominations took place after the problem had been solved. Zack notes that 'Explanations which are embedded in the flow of talk, especially in the beginning phase, may be harder to distinguish than those offered towards the end' (Zack 1994, p. 415). Two children noted that it was easier to follow explanations in the larger group of twelve than in pairs or small groups because the possibilities by that time had been narrowed down. This prompted her to ask the children to identify when helpful explanations occurred; 23 said when working in pairs, fifteen when in groups of four and 58 when in a group of twelve, thus supporting the opinion of the two peers. There was recognition of the fact that 'patterns of talk might be gender-based, with males tending to "holding the floor" presentations, and with females feeling more comfortable with collaborative "side-by-side" talk and building together' (*ibid.*, p. 415). It is suggested that greater attention needs to be paid to children's explanations in mathematical problem-solving situations in order to understand *their understanding* better and the ways in which they extend how they come to know mathematics.

Pupil perspectives

A classroom study carried out by Tomazos (1997) focused on the pupil perceptions of what goes on in the classroom. A main participant in the study was the teacher who was setting out deliberately to change her methodology, but it was the pupils' perceptions of the changes that are examined and their developing relationship with the researcher. Tomasoz argues that 'It seems that if the researcher wishes to investigate any part of the classroom milieu, there are dangers in limiting the viewpoint to the agenda of the teacher, or of the researcher, and ignoring the multiplicity of agendas assembled by the students' (*ibid.,* p. 222). The data are part of a larger case study and the issue being examined is how classroom subcultures can subvert the intentions of the teacher and how pupils' reactions 'may impinge on their teacher's efforts to change their mathematics learning environment' (p. 223). The study concerned a Year 7 class (final year of primary school in Australia) and their teacher's attempts to change her mathematics teaching to a more collaborative, group-learning situation which had proved successful in teaching English. The lessons were observed, and individual and small-group interviews were taped, and this provided the data for the study.

The excerpts from interviews with pupils indicate a degree of frankness about their teacher and show their thoughts about her feeling of insecurity in teaching mathematics, as well as her motivation. There was an implication that 'teachers ought to provide, but couldn't be expected to know, what students like' (p. 225). They considered that the teacher had tried to make mathematics 'fun'. The researcher notes that in interviews with pupils, it was a considerable time before the topic of the content of the lesson was actually touched upon. The excerpts of the lesson show how one class member offered alternative ways of approaching the logistics of collecting data from all members of the class which the teacher rather grudgingly accepted. Her main concern was that 'I need for you to have three little diagrams that indicate how frequent these colours are occurring in each person's box' (p. 225). Questions relating to procedure followed this, during which the teacher appeared to speak confidently, and thereafter the pupils carried out the task. The teacher appeared pleased with what seemed to be the collaboration and sensibleness of the pupils as they did so, and only later on interviewing the pupils did the researcher become aware of what they had really felt and thought. They clearly had not enjoyed the noise level and having to go around the class collecting data, and in the end they admitted they had 'faked' the data. Tomazos (1997) says that neither she nor the teacher had realized that the children had 'actually disengaged from the teacher's purpose and begun to follow their own interests' (p. 225). The next day, the project was continued in a more structured way but the momentum was lost and pupils began to complain that they were not learning any mathematics. They gave an overall impression of enjoyment to the teacher but were open with the researcher about their disenchantment with the situation. The pupils had 'effectively disengaged' and 'continued to operate in ways which were difficult for the teacher to interpret' (p. 228).

Evidence of this kind leads Tomazos (1997) to consider that we are too willing to focus on the teacher's agenda and the mathematics to be taught without adequately taking account of the pupils' perspective. The thrust of the argument is that we are not as aware as we should be of what these perspectives are, although they clearly affect teacher effectiveness, and she suggests that the development of theories such as enactivism (Davis 1995) which 'focuses on the fluid and complex interrelationships between organisms and their contexts, and does not privilege any one perspective over another' (*ibid.*) may be a profitable grounding to inform further research of this kind.

Children's informal reasoning

Children's informal mathematical reasoning was the focus for a study by Graves and Zack (1997) using the children's language and discursive practices as a source of data. Their theoretical background included considerations of the social and functional uses of language (e.g. Bernstein 1996, Lemke 1995, Vygotsky 1978) together with mathematics education research (e.g. Lampert 1990, Lave 1988, Schoenfeld 1992) which supports the view that children 'develop the understanding of what it is to do mathematics from the practices into which they are socialized' (Graves and Zack 1997, p. 17). The view of informal reasoning adopted is taken as argument which, for a conclusion to be valid, requires supporting evidence, and quality of argument is judged by the strength of this evidence (Voss, Perkins and Segal 1991).

The children in the study were a class of thirteen Grade 5 (Canadian) pupils and the problem focused upon was a variation of the chessboard problem, and the following description is given of the activity and its context:

> For each problem assignment, the students first work individually, then collectively in groups of 2/3, and then in groups of 4/5. Finally they meet in a large group to discuss the problem. In practical terms this means they have engaged with the problem on four separate occasions. The group discussions typically begin with a comparison of the students' answers and then proceed to a comparison of solutions and strategies. At each step of the way students are encouraged to reflect on what they did, justify their formulations with evidence, understand how someone else went about solving the problem and assess the value of strategies and approaches. (*ibid.,* p. 18)

Children's previous experience with the problem had been to work with smaller squares (4×4 and 5×5) where their strategy initially had been to count, check, then double-check. After several discussions they arrived at the strategy of collecting the squared values of the different-sized squares although it is noted that they were not necessarily aware of the nature of squared values in the form of, for example, 3^2 where they would use the number 9. Children were viedotaped as they talked in a large group over a twelve-minute period during which they sustained the discussion themselves. The transcripts presented show how the argument proceeded from one boy explicitly stating his goal to disprove a multiplication strategy used by other pupils. He uses a diagram and different values to help them to understand his argument, and is challenged. He is forced to take up the challenge and explain his thinking further and give more examples until the others agree they understand what he has done. He begins the next stage by saying 'You need to know this to understand my proof' (*ibid.,* p. 21) and after an initial statement, the discussion is taken on by three other pupils after which the presenter of the argument has to justify his claim.

In plotting the pattern of the discourse it is shown how the reasoning of the children evolves in the context of the activity as it progresses. The importance of the whole episode is that the strategies for interpreting and arguing are used specifically in the context of mathematical activity. The outcome of a study of this kind thus provides a valuable insight both into the mathematics as well as the children's ability to reason, negotiate and collaborate in generating new shared mathematical meanings.

The secondary years

In this section, the studies reflect the variations in kinds of approaches being adopted in classroom research at secondary level. They reflect similar concerns as those with respect to pupils in the primary years.

Pupils' perceptions of the purpose of a task
A study by Bell *et al.* (1997) focuses on secondary pupils' perceptions of the purposes of mathematical activities. They make the point that most metacognitive research in the past has been aimed at examining general problem-solving strategies (e.g. Lester 1988). The study they report formed part of a longitudinal study which set out to enhance the reflective activities of pupils and to provide them with learning experiences through which they could gain knowledge about specific mathematical tasks and processes. This study took part at the end of the first phase which was observational in nature and was designed:

(a) to gain information about pupils' perceptions of the purpose behind five different classroom mathematical activities; and
(b) to see how these differed from the teacher's perceived purpose.

The five lessons were taught by the same teacher to four classes (F, C, M and O) from four different schools (the ages of the pupils are not given) and the lesson topics were about multiplication and division with decimal numbers. Classes M and O had been 'exposed to a substantial number of awareness-raising interventions' while F had experienced a few such interventions and C only one. It was agreed that the teacher 'did not articulate the intended purpose of the tasks at any time, as it was the purpose of this study to discover how well the pupils could deduce this from the activities themselves' (Bell *et al.*, p. 67). Pupils were asked to describe in their own words what they thought was the intended purpose at the end of each lesson and they were also asked to rate a list of purposes on a scale of 0, 1 or 2 according to whether it was 'not a purpose', 'helped a bit' or 'was the main purpose' of a lesson (*idem*). The possible purposes were given as follows:

This lesson was to help you:

(a)   to get better at discussing and explaining;
(b)   to practise multiplying quickly and accurately;
(c)   to practise measuring and drawing accurately;
(d)   to learn how to plan and organize;
(e)   to learn when multiplying is the right thing to do;
(f)   to find the largest answer;
(g)   to get better at writing explanations;
(h)   to help you understand what multiplication really means.
    (*idem.*)

The five lessons centred on the following themes:

1.   A *concept discussion* entitled 'Believe it or not' in which pupils worked in groups of ten discussing statements such as 'Multiplying makes number bigger' and they were asked to say whether the statements were always, sometimes, or never true.
2.   *Practical construction:* 'Maximizing the volume of a box' when pupils were given a 17 cm printed square piece of paper and instructed to make a box by cutting one square cm from each corner and folding the paper, then to calculate the volume. After this, pupils were given a second piece of paper of the same size, and different groups had to cut different-sized squares from the corners to make boxes and find the volume of the results. A discussion then followed about how to maximize the volume.
3.   *A calculator investigation* – 'Maximizing a product' where pupils were given the number 11, and asked to split it three ways so that the resultant numbers, when multiplied together, would make the largest number.
4.   *Skills practice:* 'Cross Number' – In groups, pupils were given calculators and asked to complete a 'cross number' puzzle.
5.   *Recognizing the operation* – 'Do not solve it' in which groups were given 24 word problems and asked to write down the calculations that needed to be carried out for their solution. Groups were told *not* to evaluate the expressions they wrote down.

The results of pupils' responses to the 'free question' (where they could write what they felt was the purpose of a lesson) were analysed and grouped according to key phrases used. The most frequent and least frequent responses for each task are given in Table 5.1 to summarize some of the findings.

Table 5.1

Task	N	Purpose identified	%
1	91	To learn to work in a group	51
		To practise x and ÷	22
2	102	To learn about areas and/or volume	53
		To find a pattern in the results	16
3	104	To learn to use a calculator	31
		To problem solve/investigate	9
4	106	To practise x and ÷	73
		To develop mental fluency	17
5	107	To recognize when to x or ÷	29
		To solve everyday problems	10

In the tasks 1, 2 and 4, most pupils' perception of the purpose of the task was in close agreement with that of the teacher. However, in task 3 ('Maximizing a product') they found it difficult to focus on a specific purpose and were drawn to focus on the use of the calculator as a main purpose. Only two referred to an aim that had anything to do with planning and organizing how they went about solving the problem, which was the teacher's main purpose of the activity. In task 5 ('Do not solve it') many pupils simply described the task, with little analysis of what its purpose might be and 17% thought it was to improve their computation skills in spite of the fact that they were explicitly asked not to carry out any calculations.

The result of three mismatches relating to the pupils' ratings about the possible purpose of a task (as shown above) were selected because there was 'an interesting mismatch between the perceptions of the teachers and the students' (Bell *et al.* 1997, p. 70). Table 5.2 below shows the mean rating of the three perceived purposes in relation to each task that falls into this category.

Table 5.2

Purpose of task	to learn how to plan and organize	to help us understand what multiplication really means	to see how to use mathematics in our everyday lives
Task	Means of pupils' ratings (2 = main purpose, 0 = not a purpose)		
1 Believe it or not	1.05	1.20	0.37
2 Cross number	0.14	0.33	0.44
3 Practical work	0.63	0.74	0.20
4 Splitting 11	0.59	0.98	0.29
5 Do not solve it	0.40	0.94	0.78

Teachers saw the calculator investigation as having the most potential for organizing and planning, while this was rated higher in two other tasks. Task 2 was seen as helping to understand multiplication more than anything else by both teacher and pupils but pupils also saw task 3 in the same light, although in fact it consisted of fairly repetitive exercises. While both teachers and pupils rated task 5 as helping to see how to use mathematics in everyday situations, the teacher also rated task 3 (practical work) as having this as its main point, but pupils saw it as least relevant.

This kind of study highlights very clearly how a 'didactic contract' (see Schubauer-Leoni and Perret-Clermont 1997, above) can break down how the expectations of the participants can be quite different, and the kind of misunderstanding that can happen. It also echoes some of Tomazos's (1997) findings with respect to pupils' perceptions. The researchers note that the mismatch between teachers and pupils in their study was greatest where the task became more open and process-oriented, and the lowest mismatch occurred on the obvious skills practice task. After further statistical analysis of their data, the findings were related to the degree of previous experience of classes with this kind of activity. The lowest mismatch scores were found with Classes M and O who had experienced a substantial number of interventions to raise their awareness. Bell *et al.* suggest that 'the awareness-raising activities have increased understanding of the purpose of many types of lesson' and state that 'the effect is not confined to the more open or unusual styles of lesson', thus indicating some spill-over into pupils' experience in lessons not centred on problem-solving (*ibid.,* p. 72).

The effects of setting

In a study of the effects of grouping pupils for mathematics lessons, Boaler (1997) examines the pupils' perspective of the influence of the groups in which they are placed in mathematics classes in two UK comprehensive secondary schools. She notes that research in the past has focused on the inequities that arise from setting or streaming ('tracking' in the USA) for pupils who are allocated into lower sets who tend to be disadvantaged by class, race or gender (Tomlinson 1987, Abraham 1995). She argues that relatively little attention has been paid to the effects of this kind of grouping on the academic achievement of pupils. In an attempt to redress this situation, she reports two case studies which formed part of a larger study, in which quantitative and qualitative data were used to:

(a) consider pupils' responses to setted and mixed ability teaching; and
(b) consider the ways in which these types of grouping have affected the mathematical achievement of individual pupils.

The data were gathered over a period of three years when one year group in each of two schools was tracked over this period. Methodology included lesson observations, questionnaires, interviews and eliciting constructs from teachers. A wide range of assessments was also used with the pupils in the sample and both their school-based and GCSE results (an examination taken in Year 11 at the age of 16 years) were analysed. The views related to perceptions of setting were extracted from interviews with 24 Year 11 pupils from each school, although the interviews themselves were not focused specifically on setting.

The two schools for the overall study had been selected because of their differing styles in teaching the mathematics curriculum. One school (Amber Hill) taught mixed-ability classes for Years 7 and 8 but relied on setting at Year 9 (13- to 14-year-olds), a policy which had recently been enforced. This meant that the pupils concerned had experienced both 'banding' earlier on, which is described as a 'loose form of setting' and had moved to a situation of being placed in one of eight or nine sets. Teaching in Year 9 is described in the following way:

> There was no department policy that dictated the way that teachers should operate in Year 9 to 11, but all of the teachers used the same approach. They stood at the front of the class and explained methods and procedures from the blackboard for 10–20 minutes; they then set the students questions to work through in their textbooks. At periodic intervals they would stop the students and check answers, then move on to the next exercise. This process ensured that the students all moved at a similar pace through their textbooks. (Boaler 1997, p. 579)

The evidence reported is based on the interviews with twenty pupils of the 24, taken from the top four sets, a factor dictated by the aims of the study as a whole; the remaining four pupils were from Set 7.

The second school (Phoenix Park) had a similar catchment area (mainly white, working-class parents) to the first and there were no significant differences in terms of social class, ethnicity, gender or ability. The pupils had been taught in Years 7 and 8 using the same scheme as the first school, which was based on individualized work cards where pupils could work at their own pace. The teaching approach is described in the following way:

> In mathematics lessons at Phoenix Park the students work on open-ended projects at all times – the department does not use any textbooks or published materials. The teachers generally give the students open starting points, such as 'What it the maximum sized fence that can be built out of 36 gates?' The students are then encouraged to go away and work on this information, develop questions of their own, extend the problem and use mathematics to answer their questions, for approximately three weeks. At the end of this time they complete a description of the activities they have worked on and their results. (*ibid.*, p. 580)

The results of interviews with pupils at Amber Hill showed considerable disaffection with mathematics, and in the interviews it emerged that most of this was caused by the effects of setting. The reasons for this are identified as follows:

1. *Pacing* – Pupils expressed a dislike for the loss of control over the pace at which they worked as a result of the setting procedure. Boaler suggests that in a setted situation, the teacher pitches the lesson at the 'middle' of a group believing that the faster or the slower pupils will be able to adjust their pace accordingly. However, this proved not to be the case and many pupils in both the higher and the lower sets who could not make the adjustment became disaffected and began to underachieve. Girls seem to find the pace too fast, while it was the boys who generally found it too slow. Boaler points out that this reveals a limitation of a class-taught approach generally and that 'it shows how difficult it is to teach a group at the same pace, even when they are supposed to be of "homogeneous" ability' (*ibid.*, p. 582).

2. *Pressure and anxiety* – No significant difference in attitudes was found between boys and girls at Phoenix. However, mathematical anxiety was found to be prevalent at Amber Hill, particularly among the girls. The girls linked this anxiety 'not to the intrinsic nature of mathematics but to the pressure created by setted classes' some of which arose from the pace set by the teacher (p. 584). Another facet of pupils' anxiety related to what were perceived to be competitive standards set up by being within a particular group which they felt they had to live up to and the pressure of constant peer judgement they felt they underwent. Boaler notes that this is contrary to the notion that competition raises achievement, and her results suggest that the pressure arising from this aspect of setting was felt by the most able as well as pupils in lower sets.

3. *Restricted opportunities* – The results of analysis of pupils' achievement over the three years showed a marked decline in the achievement of pupils at Amber Hill. Those in lower sets felt 'cheated' because they knew the system is such that the highest grade they could achieve was a C or, in the lowest sets, only a D would be available to them (as a result of the tiered examination system in the UK). They were aware that even if they were to get 100% on an examination they would still only be awarded a D grade, and Boaler notes that 'The students believe they had been restricted, unfairly and harmfully, by their placement into sets' which limited their opportunities generally' (p. 586).

4. *Setting decisions* – Many pupils considered the sets in which they had been placed did not reflect their true ability. Boys in particular felt they had been placed in a set because of their behaviour and not because of their mathematical potential.

Statistical analysis of pupil achievement in both schools showed two interesting outcomes:

(a) Because of the supposed benefits of setting, a high correlation might be expected between the ability test given at the end of Year 8 and the Year 11 exams, for pupils of Amber Hill, while a smaller correlation between the two might be expected at Phoenix Park. However, the statistical analysis indicated that the correlation was considerably higher for Phoenix Park than Amber Hill (0.67 and 0.48 respectively).

(b) There was a correlation between social class and set in which a pupil was placed. At Amber Hill the results suggested that pupils from a lower social class were likely to be placed in low sets. In contrast, a negative correlation showed that there was a small tendency for students of a lower social class to be placed into a higher examination group at Phoenix Park (*ibid.*, p. 588).

Although the extract reported here is based on interviews with only 24 pupils mainly from one school, the pupils' perspectives show considerable dissonance with the intended advantages of a restrictive system of setting according to ability. A more extended study of mixed-ability grouping by Linchevski and Kutscher (1998) supports Boaler's evidence with respect to its effectiveness. Having compared the achievement of pupils in classes of same-ability setting with those of mixed-ability setting, they concluded that this form of organization was 'not detrimental to their achievements' (p. 551).

Pupils in a multimedia classroom

Nemirovsky *et al.* (1998) build on earlier work related to the use of technology in the mathematics classroom (Kaput and Roschelle 1997) in which they describe the question they explore in their work as follows: 'We address the question of how we might exploit the interactive technologies to democratize access to ideas that have historically required extensive algebraic prerequisites' (p. 195). The focus is to use technology for the *generation* of phenomena and not merely *modelling* phenomena which allows pupils to embody their intentions and theories, and both to interact with them and to control them at the same time. The example is given of how pupils 'can use mathematical notations or other controls within a computer environment to control the motion of physical "Minicars" that move on tracks. The same means can drive a simulation "within" the computer.'

The purpose of the activity described is to combine both types of capabilities to generate phenomena. The following distinctions are drawn:

1. Phenomena 'inside the computer' vs physical phenomena 'outside the computer.
2. 'Target phenomena' taken as the subject of mathematical description or 'control (e.g. motion) vs 'notational phenomena' embodied within notations that may be used to describe or control the former. (*ibid.*, p. 288)

They note that the notions of 'targets' and notations are being used in an heuristic way in order to expose the structures of activities, which involves co-ordination across more than one realm of phenomena. They note the development of Microcomputer Based Laboratory (MBL) devices to enrich work with simulations, creating new situations such as parades or dances that 'embed function-descriptions of motion in interestingly complex relations' (*ibid.*, p. 290). Devices have been designed (Nemirovsky 1993) to expand the realm of physical action to allow the pupils to control air flow, rotary motion or the shape of a surface, 'either by directly acting on the physical device, *or by using a mathematical function defined on a computer that "drives" the device*' (*idem*, author's emphasis). This kind of function has been labelled 'Lines Become Motion' LBM). Using this phenomena-generating capability results in a situation where 'the mathematical notation that controls a phenomenon also models it' and the pupil's intentions are made visible and explicit, and are testable through the pupil's own control of the phenomena.

A classroom situation is described in which a Grade 9 algebra class was involved with their teacher in experimental activities illustrating interactions among physical, cybernetic and notational spheres of activity. In other words, observable action, computers and mathematical representation were being brought together in a single situation. The action was a toy car going down an inclined plane. The car was connected to a computer that showed its velocity in 'real time', and an LBM 'Minicar' whose motion could be controlled by a graph drawn on the same computer screen. The activity centred around the teacher's proposal to 'measure the speed of the car on the inclined plane to generate a graph that would correspond to the downhill part' (*ibid.*, p. 291). During the activity, pupils noted a 'bump' in the graph where there was a 'negative velocity peak' and explained it in terms such as 'it hit something' or 'it bounced back'. They eventually arrived at the meaning of negative velocity as 'going backwards' (*ibid.*, p. 292). This observation was built upon and then developed throughout the session by exploring connections between 'small bumps' on the graph with the action of the car as well as the pitch of the sound with the speed of the car. The pupils reached a stage, with guidance from the teacher, where they inverted a graph of a situation involving a sled going down a hill, and were able to reach an understanding of the beginning and end segments of the graph:

> *Jesse:* When the graph is going up, the sled is going down. When the graph is going down, the sled is going up or something. That's confusing, because ...
> *Ben:* Just think about it as the speed going up, not the sled going up. We are talking about the speed and the time, not the sled. It has to do with the hills because it has to do with what part of the hills you are on, but it is really about speed. (*ibid.*, pp. 292–3)

The research question was concerned with exploring how technologies and learning environments change the nature of pupils' mathematical learning of this kind by tapping into their cognitive, linguistic and kinaesthetic resources. Analysis of the results suggested the interplay of three aspects of the episode described:

1. *Incorporating the kinaesthetic activity* – The action of the car rolling down a 'hill' introduced the pupils to the notion of 'negative velocity' and its depiction as a graph.
2. *Empowering notations to create phenomena* – The interplay between controlling the car by hand and seeing the 'bumps' in the graph gave the graph a new status of being able to create phenomena.
3. *Experimenting with similarities and differences between virtual and physical phenomena* – The pupils were involved in 'imaginary' simulations with sleds going up and down a hill; they also engaged with some of the differences between cybernetic and physical phenomena through consideration of starting with a 'zero velocity'.

The conclusion is drawn that 'these tools and learning environments help to recruit linguistic and cognitive resources from their everyday experience' (*ibid.*, p. 294). By relating observed physical action (the car on the inclined plane), the pupils were able, through the computer, to 'see' this graphically represented and to extend their understanding of one situation to others of a similar kind which had been part of their personal experience.

TEACHER PERSPECTIVES

The studies that follow in this section focus on the 'teacher in action' which reflects the holistic approach of much current study in the field of mathematics education.

Teachers in the primary years

In a study of teachers' informal assessment in the classroom, Watson (1998) explores the way in which teachers make judgements and either adjust or do not adjust these judgements as teaching progresses. She reviews seven incidents of teachers' informal assessments, five of which are made in the primary school context. She identifies four main ways this kind of judgement made by teachers is important:

1. pupils' self-image with respect to mathematics is, at least in part, the result of their perceptions of how the teacher sees them;
2. teachers make judgements about how to respond to pupils in a one-to-one situation;
3. the curriculum followed by pupils is determined by the teacher (in the UK setting) usually based on teacher judgements alone;
4. teachers' assessments (in the UK) are a formal part of 'high stakes assessment' which affect pupils' later education and social opportunities. (*ibid.*, p. 170)

Teachers who form the focus of the study were observed in their normal classroom situation (upper primary and lower secondary) and were also interviewed. Long-term observations of ten pupils aged 10 to 12 years were also undertaken. The selected incidents were chosen because some 'exceptional mathematical performance is displayed which the teacher notices and talks about' (pp. 170–1). The features of these incidents are examined in order to answer the following question:

> What factors influence the teacher to accept a single mathematical performance as noteworthy evidence to add to her picture of the student's achievement, and what leads her to dismiss the performance as adding nothing? (p. 171)

The incidents are described and features identified for each incident relating to the decision made by the teacher during the observation. The two examples that follow give an idea of the kind of incidents under consideration.

> *Incident 1.* During a whole class lesson an 11-year-old pupil had said nothing. In the following lesson, she carried out an operation correctly and although being challenged by the rest of her group, stuck to her view about its correctness. The teacher described this as evidence of the pupil being 'a strong mathematician' and throughout the term, held to this view although there was no further evidence of any notable mathematical ability. (p. 171)

> *Incident 2.* A 10-year-old pupil dealing with flowcharts wanted to know what 'inverse' meant and was told by the teacher to look it up in a mathematical dictionary. She did this and constructed a meaning for inverse which she used successfully with the flowcharts. The teacher judged her to be a 'strong mathematician because she could think logically and had strategies to use when she did not understand something'. (p. 172)

Features are identified for each incident and related to the decision made by the teacher during the observation. The analysis of these incidents reveals 'the social skills of pupils able to draw the teacher's attention to their thinking and were *a priori* exceptional when compared to the performance of the rest of the class, the teacher's expectations for the class or the teacher's prior impressions of the pupils' (*ibid.*, p. 175). While some teachers may be willing to alter their view of a pupil on the basis of one incident, others will do so over time and with further evidence. However, there were teachers who did not alter their view even after an accumulation of evidence. Watson acknowledges that although this may be due in part to the teacher's view both of mathematics and what constitutes exceptional performance, there may be other factors contributing to this kind of lack of adaptability of judgement: 'The teacher's own self-esteem may depend on preserving her confidence in making judgements, her professional functioning (such as discriminatory curriculum practices) might depend on making hasty judgements and ignoring contrary evidence' (*ibid.*, p. 175). Only two examples are given here but others reported indicate a range of teacher behaviour showing a similar disinclination to change their views even when different evidence arises. This may result in potentially able mathematics pupils being ignored, or conversely, weaker pupils not being given the support they may need because the teacher is convinced they are stronger than they are, either because of their social skills or occasional successes.

Teachers' beliefs and practice

Groves and Doig (1998) build on previous work (e.g. Doig 1997) in exploring teachers' beliefs and how these relate to what they do in the classroom, particularly in relation to the nature and role of discussion. In previous studies, they found that 'many teachers, regardless of their stated theoretical frameworks for learning and teaching, were uncomfortable with whole-class discussions based on the notion of scientific dialogue' (Groves and Doig 1998 p. 17). Scientific dialogue is viewed here as classroom discourse that can be progressive in the same sense as science, 'with the generation of new understandings' (Bereiter 1994) which requires the commitment of those involved to work towards establishing ideas that can be tried and tested. They refer to the work of Cooney and Shealy (1997) which emphasizes the importance of teacher beliefs and an understanding of how they are structured in order to bring about change in teacher education that is not essentially haphazard.

The study involved the teachers of a Grade 1, 4 and 5 class. Data were gathered from interviews with each of the three teachers about their beliefs relating to the teaching and learning of mathematics and they were videoed teaching their classes. Descriptions are given of each of the teachers and a summary of what took place in the classroom sessions observed. The data indicate that all three teachers 'showed remarkable consistency between their beliefs and practices' (*ibid.*, p. 23). However, analysis of the interaction within the classrooms revealed that two of the three were actually engaging in practice that was not conducive to the achievement of their stated goals, and in some instances, was antithetical to them. Although enhancing children's self-esteem was seen to be an important function of involving them in discussion and respecting their views, there were few genuine instances when this happened in the case of two of the classrooms. Groves and Doig conclude that for the effective development of teacher practice to happen, a first priority must be to problematize some aspects of current practice such as the relationship between teacher intentions and classroom outcomes.

Effective teachers of numeracy

Another study concerned with teachers' beliefs and practices is reported by Askew *et al.* (1998). They give the outcomes of one aspect of their work relating to the effective teaching of numeracy in which they carried out case studies of eighteen teachers to explore their beliefs about:

(a) what it means to be numerate;
(b) how pupils become numerate; and
(c) the roles of teachers.

They draw on research which indicates that teachers have difficulty in adopting new practice when they do not understand the principles which underlie it (Alexander 1992) and that while they may invoke the notions of 'good practice', they do so without actually carrying out the practices which are entailed (Desforges and Cockburn 1987).

The eighteen teachers were identified as effective teachers of numeracy on the basis of evidence from the respective schools, and on a test especially designed to test numeracy. Average gains by pupils were used to classify teachers into three categories according to their effectiveness: *highly effective*, *effective* or *moderately effective*. The case study teachers were evenly grouped across the 5- to 11-year age range and data were gathered over two terms through classroom observations and interviews. Classroom observations focused on the following:

(a) organizational and management strategies;
(b) teaching styles;
(c) teaching resources;
(d) pupils' responses.

Three different types of interview were carried out with the teacher. The purpose of the background interview was to elicit information about experience and training, and about practices and beliefs in relation to numeracy, what they thought their effectiveness in relation to teaching numeracy was due to, and why. The concept mapping interview centred on a task designed to gain an insight into the teacher's understanding of mathematical processes related to numeracy. The third was a 'personal construct' interview which focused on the group of pupils currently being taught by the teacher, in which the teacher's knowledge and beliefs about their pupils and how they had become numerate were explored.

Analysis of the data led to the identification of three models of beliefs that appear to be important in understanding teachers' approaches to the teaching of numeracy:

Connectionist – beliefs based around both valuing pupils' methods and teaching strategies with an emphasis on establishing connections within mathematics;

Transmission – beliefs based around the primacy of teaching and a view of mathematics as a collection of separate routines and procedures; *Discovery* – beliefs clustered around the primacy of learning and a view of mathematics as being discovered by pupils. (Askew *et al.* 1998, p. 27)

There were identifiable links between orientation and practice which are reported under the following headings:

- *The role and nature of mental strategies* – All three saw number bonds and multiplication facts as a baseline of expectations with respect to numeracy, although connectionists saw this extending beyond number facts to using mental strategies in a flexible way in order to develop competency.
- *Teacher expectations* – Connectionists emphasized challenging pupils of all abilities, while transmission- and discovery-orientated teachers approached the curriculum differently for the low-achieving children.
- *Interaction* – There was a high degree of discussion between teacher and whole class, teacher and pupil groups, teacher and individual pupils, pupil and pupil in the connectionist teachers' classes. In one of the most effective schools, there was an expectation that pupils would *explain* their thinking from the age of 5 years.
- *Role of mathematical application* – The discovery and transmission teachers viewed this as pupils putting what had been learned into context with selecting what was to be used from the mathematics they already knew, as the only challenge. The connectionist teachers saw application as offering the potential for pupils to learn new concepts.

Askew *et al.* conclude that 'The importance of these orientations lies in how practices, while appearing similar, may have different purposes and outcomes depending upon differences in intentions behind these practices' (*ibid.*, p. 32). They suggest that 'the pedagogical purposes behind particular classroom practices may be as important as the practices themselves in determining effectiveness' and that 'these orientations need to be explicitly examined in order to understand why practices that have surface similarities may result in different learner outcomes' (*idem*).

Teachers in the secondary years

In a study of the role of talk in a secondary mathematics classroom, Adler (1997) explores the visibility and invisibility of language in the context of mathematical teaching and learning. The work was carried out in a multilingual classroom in South Africa where the medium of teaching was English but there were pupils whose first language was Sesotho and Zulu. The focus of the study is the notion of language as a tool and a resource, and it is seen as enhancing mathematical learning insofar as it makes mathematics accessible to pupils in a teaching–learning situation. Thus it takes on a particularly crucial role in a multilingual classroom. Adler uses the idea of *transparency* together with *visibility* and *invisibility* as main aspects of the theoretical background informing her work, and cites Lave and Wenger (1991) as follows:

> invisibility in the form of unproblematic interpretation and integration (of the artifact) into activity, and visibility in the form of extended access to information. This is not a simple dichotomous distinction, since these two crucial characteristics are in a complex interplay. (p. 102)

The language through which mathematics is mediated should not intrude to the extent where it masks the mathematics that is being taught and thus denies pupils access to it. Adler (1997) suggests that a socio-cultural perspective such as that of Vysotsky (1962) provides the conceptual tools with which to understand and explain the nature of classroom learning and notes that,

> From this sociocultural perspective, the teaching and learning of mathematics in multilingual contexts needs to be understood as three-dimensional. It is not simply about access to the language of learning (in this case English). It is also about access to the language of mathematics (educated discourse and scientific concepts) and access to classroom cultural processes (educational discourse). (p. 3)

The research reported was part of a wider study in which in-depth initial interviews, classroom observations, reflective interviews and workshops formed the methodology. The episode in the study is a workshop in which a teacher is presenting a videotape of one of her classes. Adler describes her as a teacher whose classes while 'largely teacher-directed, are interactive and task-based' (Adler 1997, p. 4), and in the session shown, she is teaching trigonometry to a Grade 10 class. The problem others are invited to discuss with her is 'whether or not explicit language teaching actually helps, or whether and how working on pupils' ability to talk mathematics is a good thing and "saying it" is indicative of understanding, of knowing'. She has tried to develop mathematical language explicitly as part of teaching in her multilingual classroom. She realizes, on considering the video with others, that she had gone on too long in the lesson in trying to lead the pupils to state clearly what they thought trigonometry was all about and to use the language appropriately and accurately. The pupils' responses indicate confusion, and through a process of scaffolding using questions, she tries to get them to focus on the misuse of the word *independent* to the point where 'explicit language teaching is a struggle' (*ibid.*, p. 7). The language loses its role as a medium and itself becomes the message (Pimm 1996).

The dilemma posed in a situation like this is crucial:

> If Helen 'goes on too long', she diminishes pupils' opportunities to use educational discourse and inadvertently obscures the mathematics at

play. If she leaves too much implicit then she runs the risk of losing or alienating those who most need opportunity for access to educated discourse. (*ibid.*, p. 7)

Adler notes that the important factor with respect to visibility and invisibility of this kind and in this context, is its management, and concludes: 'There is no resolution of the dilemma of transparency for mathematics teachers, only its management through careful mediational moves when making talk visible in moments of practice' (*idem*). In other words, teachers constantly have to make professional judgements about the prominence given to mathematical language, as to any other tool used in the classroom, in order that pupils learn mathematics more effectively.

Characterizing mathematics teaching in the classroom

Jaworski and Potari (1998) use the concept of the Teaching Triad in their research in which the interactions within a mathematics classroom and the teachers' thinking which motivates the lessons are studied. They note that this involves analysis at the *micro* level with respect to the interactions that take place and at a *macro* level with the analysis of the reflections and thinking of all the participants. The Teaching Triad (Jaworski 1991) is a framework developed in order to help to identify the many factors that characterize the teaching in a mathematics classroom. The three elements of the triad are:

1. *Management of Learning* (ML) which provides a description of the teacher's role in the classroom as it is constituted by the teacher and pupil.
2. *Sensitivity to Students* (SS) which concerns the teacher's knowledge of pupils and the attention to their needs.
3. *Mathematical Challenge* (MC) describes the challenges that are presented to pupils to bring about mathematical thinking and activity.

All three of the elements are closely interlinked with each other and are seen as interdependent.

The researchers worked with two teachers who had participated in an earlier study with them which focused on investigative teaching. For this study, a number of lessons were designed for classes with pupils aged 11 to 17 years, and were observed by one or both researchers. It is noted that 'For the teachers, a major objective was to gain insights into their own practices in order to develop or improve their teaching. Researchers sought the nature of the practices observed, issues arising and theoretical conceptualizations of the teaching' (*ibid.,* p. 89).

They cite Cooney (1994) who suggests that '. . . we should be interested in how local theories about teachers can contribute to a more general theory about teacher education' (p. 627). With this in mind, they use the Teaching Triad as a construct in such theorizing about the complexity of practices in the classroom teaching of mathematics. Two episodes are given which focus, first, on the teacher as she introduces a lesson; and second, on the teacher interacting with a group of pupils. The episodes were analysed according to the elements of the triad. The analysis shows how the interaction has been coded and one such episode produced the following results:

- *Management of learning* – evident in (a) the environment created (grouping of pupils, the task itself and the expectations with respect to pupils' collaborative working, mathematical thinking and autonomy in their activity); (b) interaction of teacher with individuals and groups, (focusing questions to help pupils develop and articulate concepts, maintaining balance between teacher input and pupil autonomy); (c) reminding pupils of classroom norms and what is required by the task.
- *Sensitivity to students* – shown by (a) attention to pupils as individuals or in groups; (b) discovering what they have done; (c) trying to find out what they are thinking; (d) praising achievements and judging results of their activity in the light of objectives that had been set.
- *Mathematical challenge* – implicit in (a) the task; (b) the expectation that pupils direct their own activity; and (c) the focused questions to help in coping with the concepts they have to deal with.

Jaworski and Potari identify the cognitive aspect of the mathematical challenge and note that sensitivity appears to be both in the affective and the cognitive domains. The teacher was clearly concerned throughout about the affective elements in her classroom and showed an awareness of pupils' self-esteem in her exchanges with them. They suggest that 'An analysis which attends to cognitive and affective elements and their relation to sensitivity and challenge in the teaching triad has potential to be informative to teaching development'. Reflection on what had taken place helped the teacher 'analyse further how she conceived her sensitivity both in planning her lesson and in reflecting on it' (p. 94). In conclusion the researchers suggest that analysis of this kind can show patterns in interactions which can lead to propositions about such interactions which can, in turn, be tried and tested by other teachers. In particular, the cognitive and affective interface existing within the triad leads to further insights about what makes teacher–pupil interaction effective.

IMPLICATIONS FOR TEACHING

Current classroom research in mathematics education has, at its centre, the nature of mathematics as a subject and the influence it exerts on the context in which it is taught and learned. Perceptions of mathematics as a discipline have changed among a large section of the teaching and research community, and these perceptions, in turn, influence the kinds of questions being addressed in the body of research. Some of the main implications for teaching that can be drawn from the studies considered in this chaper follow below, but it is clear that there are many other messages that are identifiable by the reader which may have equal potential for affecting practice.

Theoretical issues

1. There is a need for greater awareness of how mathematics as a subject can affect the classroom. The idea of a culture of the mathematics classroom is not merely a metaphor but a phenomenon that needs to be attended to. Pupils and teachers bring their own perceptions of mathematics to the classroom, as well as beliefs and values in relation to learning and applying mathematics. For example, a pupil may dislike the subject on entry to school because their parents did not like it, or a teacher may dislike teaching it because of their own lack of confidence in mathematics. All such matters affect the quality and nature of the interaction in the classroom and ultimately the effectiveness of the teaching and learning that takes place in it.

2. The swing to more descriptive, qualitative research that is interpretative rather than predictive is likely to be more accessible to teachers. Teachers may find it more relevant and identify with it and 'see themselves in it'. Increasingly as equal partners in participating in research, they will influence what is researched and how it is done and their own theories will play a more prominent role. This aspect of their professionalism needs to be acknowledged more widely in order to increase participation at this level.

3. Theories such as the *didactic contract* are helpful frameworks against which to examine and even judge what goes on in the classroom. It helps to make clear the expectations of teachers and pupils and can give a structured overview of the pattern of interchange that is set. The example of the study in this chapter is a reminder of how pupils perceive us as mathematics teachers, and of some of the unintended but undesirable effects our reactions to some classroom situations may have, for example with the unconscious differentiation that can happen as a result of the judgements made on superficial criteria.

4. All of the theoretical perspectives emphasize the *social* character of mathematical learning and the importance of the interaction within the classroom as a mutually constructive situation where pupils learn both from the teacher and their peers. The teacher creates an environment in which this can happen and *intervenes* as an active participant, and it is the nature and timing of that intervention that is crucial to pupils' effective learning.

Pupils' perspectives

1. The importance of talk in the mathematics classroom is heavily emphasized in the research, and it is talk of a variety of kinds. There is the talk that involves pupils having to *explain,* and *reason* about, their

thinking, either to the teacher or to peers, as well as group talk in which they have to *discuss, argue* and *justify* their thinking. As they talk about the mathematics they *clarify* their thinking and become aware of the importance of alternative *strategies* as they learn from others.

2. Group talk in which pupils listen to each other's explanations and ideas can establish mutual respect, an interdependence and a sense of responsibility that emphasizes the importance of the social character of mathematics as well as of the learning process. This needs to be emphasized in mathematics to try to overcome the image of the subject as detached from everyday life. The image of mathematics as a *social tool* is a powerful one and pupils need this tool to function effectively in a technological world. The social benefits of group discussion also have a spin-off effect for mathematics. Pupils who gain in self-confidence through open discussion of mathematics may come to see mathematics itself in a less threatening light. However, it is also important to remember the possibility of gender effects and to ensure that both girls and boys participate in such discussion.

3. It is important to be aware that pupils' perceptions of what is intended in a mathematics lesson may not be the same as the teacher's. It would seem that with preconceived ideas about mathematics in particular, it is easy for pupils to assume what the teacher's intentions in a lesson are, when in fact it can be something quite different, and it may be too easy for pupils to assume what they see as the obvious (e.g. as in the case of investigative work with a calculator which they associated with staightforward calculation). Sharing with pupils the reason *why* something is being done may in itself seem obvious and therefore may not always happen. The intention can become so embedded in the way a lesson is structured or in the nature of the activity, that it can easily assume secondary status, or worse, be lost.

4. Part of pupils' misconceptions about the purpose of a lesson may be due to the fact they may have to switch from one mode of teaching to another with which they have had little experience. If they are accustomed to being taught almost entirely through textbooks, they may find more open-ended, problem-solving situations difficult to cope with. This highlights the important need for balance in methodologies from the pupil's point of view so that there is not a dissonance between their expectations and the pedagogy.

5. The evidence points to the fact that a problem-solving approach has clear benefits for pupils in helping them to approach mathematical problems of all kinds in a more structured way. Practice in identifying the main features of a problem and rejecting redundant information, and looking for relationships and strategies in a problem-solving situation, are all transferable skills that can be used in all areas of mathematics as well as in other subjects.

6. While individual teachers may not have much control over the grouping in which pupils are taught in their school or classrooms, an awareness of the

possible unlooked-for effects of teaching groups is very important. Pupils can clearly receive messages that are demotivating, and their self-confidence and self-image suffer accordingly. The indications that a mathematics teaching–learning situation can be associated with competitiveness suggests adding an extra layer of difficulty which teachers have to penetrate in order to ensure that all their pupils have reasonable access to mathematical learning.

Teacher perspectives

1. There is a constant need to be reflective about the judgements made in the classroom with respect to pupils and their ability, and the evidence on which they are based. The effects of first impressions can be very strong, but flexibility is needed to overcome these when relevant evidence presents itself. The research indicates that often we are unconscious of the basis for our judgements, and by having to justify them to ourselves through reflection, we are forced to identify reasons for thinking and acting the way we do with respect to pupil potential and performance.

2. Teachers have beliefs about mathematics, and the evidence shows that these beliefs affect how they teach the subject. The methodology in itself conveys powerful meanings to pupils, and the self-reflectiveness of teachers referred to above is also needed with respect to the methodologies used. The evidence indicates that teachers may *think* they are teaching in one mode but actually are performing in another. If we believe our methodology is of a particular kind that has a set of intended outcomes but in reality it is *not* that methodology, then clearly the intended outcomes will not be achieved. If, for example, being able to reason mathematically is an intended goal and, at the same time we do not give pupils the opportunity to explain their thinking, then we are not engaging in the pedagogy appropriate to achieving that goal.

3. The increased importance placed on talking in the mathematics classroom calls for balance in another area. While it is important for pupils eventually to acquire correct mathematical terminology and to learn what it is to define something, there is once again a need for flexibility in how this is approached. This may be one of the areas where sensitivity in exchanges with pupils is particularly important, where explanations using their own language are seen to be valued and the arrival at the accepted terminology is reached over time.

4. There is evidence that using technology in the mathematics classroom need not make it an antisocial situation. The potential it holds for learning in which mathematics is both generated and shared at the same time, seems great. With effective software, the talk that emerges necessitates children identifying mathematical ideas and phenomena in their own language and can take them beyond the mathematics of a scheme or a textbook. This is a considerable challenge for teachers, but it also offers a

shared learning situation for teacher and pupils. Teachers need to have familiarized themselves with a program being used and to try to anticipate the potential critical learning ideas in it, but they also have to expect the unexpected and let pupils' mathematical thinking evolve.

Bibliography

Abraham, J. (1995) *Divide and School: Gender and Class Dynamics in Comprehensive Education.* London: Falmer Press.

Adler, J. (1997) The dilemma of transparency: seeing and seeing through talk in the mathematics classroom. *Proceedings of the 21st Conference for the International Group for the Psychology of Mathematics Education.* 2, 1–8.

Ainley, J. (1995) Re-viewing graphing: traditional and intuitive approaches. *For the Learning of Mathematics.* 15.2, 10–16.

Ainley, J. and Pratt, D. (1995a) Mechanical glue?: Children's perceptions of geometric construction. *Proceedings of the Third British Congress of Mathematics Education.* 97–102.

Ainley, J. and Pratt, D. (1995b) Planning for portability. In Burton, L. and Jaworski, B. (eds) *Technology and Mathematics Teaching. A Bridge Between Teaching and Learning.* Bromley: Chartwell Bratt, 433–48.

Ainley, J., Nardi, E. and Pratt, D. (1997) Making connections through active graphing. *Proceedings of the Day Conference of the British Society for Research into Learning Mathematics, held at the University of Nottingham.* 1–7.

Alexander, R. (1992) *Policy and Practice in Primary Education.* London: Routledge.

Alseth, B. (1998) Children's perceptions of multiplicative structure in diagrams. *Proceedings of the 22nd Conference of the International Group for the Psychology of Mathematics Education.* 2, 9–16.

Alvarez, M. E. (1994) Various representations of the fraction through a case study. *Proceedings of the 18th Conference of the International Group for the Psychology of Mathematics Education.* 2, 16–23.

Amir, G. and Williams, J. (1994) The influence of children's culture on their probabalistic thinking. *Proceedings of the Eighteenth International Congress of the Group for the Psychology of Mathematics Education.* II, 24–31.

Anghileri, J. (1995) Division and the role of language in the development of solution strategies. *Proceedings of the Third British Congress of Mathematics Education (Part 1).* 113–20.

Anghilieri, J. and Baron, S. (1997) Working with 3-D shapes – poleidoblocs. *Proceedings for the Day Conference held at University of Nottingham, Saturday 1st March 1997.* 12–17.

Arnon, I., Dubinsky, E. and Nesher, P. (1994) Actions which can be performed in the learner's imagination: the case of multiplication by an integer. *Proceedings of the 18th Conference of the International Group for the Psychology of Mathematics Education.* 2, 32–9.

Arzarello, F., Bazzini, L. and Chiappini G. (1994) The process of naming in algebraic problem-solving. *Proceedings of the 18th International Conference for the Psychology of Mathematics Education.* II, 40–7.

Askew, M., Brown, M., Rhodes, V., Wiliam, D. and Johnson, D., (1997) Effective teachers of numeracy in UK primary schools: teachers' beliefs, practices and pupils' learning. *Proceedings of the 21st Conference of the International Group for the Psychology of Mathematics Education.* 2, 25–32.

Assessment of Performance Unit (1980) *Mathematical Development: Secondary Survey, Report Number 1.* London: HMSO.

Aubrey, C. (1993) An investigation of the mathematical knowledge and competencies which young children bring into the school. *British Educational Research Journal.* 19.1, 27–41.

Baker, D. (1995) Numeracy as a social practice: primary school numeracy practices. *Proceedings of the Third British Congress of Mathematics Education (Part 1).* 128–35.

Baroody, A. J. and Ginsburg, H. P. (1986) The relationship between initial meaningful and mechanical knowledge of arithmetic. In Hiebert, J. (ed.) *Conceptual Knowledge and Procedural Knowledge: The Case of Mathematics.* Hillsdale, NJ: Lawrence Erlbaum Associates. 75–112.

Baroody, A. J. and Hume, J. (1991) Meaningful mathematics instruction: the case of fractions. *Remedial and Special Education.* 12, 54–86.

Bartolini Bussi, M. G. (1991) Social interaction and mathematical knowledge. *Proceedings of the Fifteenth Annual Meeting of the International Group for Mathematics Education.* 1, 1–16.

Batanero, C., Estepa, A., Godino, J. D. and Green, D. R. (1996) Intuitive strategies and preconceptions about association in contingency tables. *Journal for Research in Mathematics Education.* 27, 151–69.

Batanero, C., Godino, J. and Estepa, A. (1998) Building the meaning of statistical association through data analysis activities. *Proceedings of the 22nd Conference of the International Group for the Psychology of Mathematics Education.* 1, 221–36.

Battista, M. T. (1990) Spatial visualization and gender differences in high school geometry. *Journal for Research in Mathematics Education.* 21, 47–60.

Battista, M. T., Clements, D. H., Arnoff, J., Battista, K. and van Aiken Burrow, C. (1998) Students' spatial structuring of 2D arrays of squares. *Journal of Research into Mathematics Education.* 29, 503–32.

Bednarz, N. and Dufour-Janvier, B. (1994) The emergence and development of algebra in a problem-solving context: a problem analysis. *Proceedings of the 18th Conference of the International Group for the Psychology of Mathematics Education.* II, 64–71.

Behr, M. J., Harel, G., Post, T. and Lesh, R. (1992) Rational number, ratio and proportion. In Grouws, D. (ed.) *Handbook of Research on Mathematics Teaching and Learning.* New York: Macmillan Publishing Company. 296–333.

Beishuizen, M. (1997) Mental arithmetic: mental recall or mental strategies. *Mathematics Teaching.* 160, 16–19.

Bell, A., Phillips, R., Shannon, A. and Swan, M. (1997) Students' perceptions of the purposes of mathematical actitivities. *Proceedings of the 21st Conference of the International Group for the Psychology of Mathematics Education.* 2, 65–72.

Ben-Chaim, D, Fey, J., Fitzgerald W. M., Benedetto, C. and Miller, J. (1998) Proportional reasoning among 7th Grade students with different cultural experiences. *Educational Studies in Mathematics.* 36, 247–73.

Ben-Chaim, D., Lappan, D. and Houang, R. T. (1988) The effect of instruction on spatial visualization skills of middle school boys and girls. *American Educational Research Journal.* 25, 51–71.

Bereiter, C. (1994) Implications of postmodernism for science, or, science as a progressive discourse. *Educational Psychologist.* 29, 3–12.

Berger, P. L. and Luckman, T. (1966) *The Social Construction of Reality.* Middlesex, England: Penguin Books Ltd.

Bernstein, B. (1996) *Pedagogy, Symbolic Control and Identity.* London: Taylor and Francis.

Biggs, J. B. and Collis, K. F. (1982) *Evaluating the Quality of Learning: The SOLO Taxonomy.* New York: Academic Press.

Bills, L. (1997) Stereotypes of literal symbol use in senior algebra. *Proceedings of the 21st Conference of the International Group for the Psychology of Mathematics Education.* 2, 73–80.

Bloedy-Vinner, H. (1994) The analgebraic mode of thinking – the case of parameter. *Proceedings of the 18th Conference of the International Group for the Psychology of Mathematics Education.* II, 88–95.

Boaler, J. (1993) The role of contexts in the mathematics classroom: do they make mathematics more 'real'? *For the Learning of Mathematics.* 13.2, 12–17.

Boaler, J. (1997) Setting, social class and survival of the quickest. *British Educational Research Journal.* 23, 575–95.

Boero, P., Dapueto, C., Ferrari, P., Garuti, R., Lemut, E., Parenti, L. and Scali, E. (1994) Aspects of the mathematics-culture relationship in mathematics teaching–learning in compulsory school. *Proceedings of the 18th Annual Meeting of the International Group for the Psychology of Mathematics Education.* 1, 151–66.

Booker, G. (1998) Children's construction of initial fraction concepts. *Proceedings of the 22nd Conference of the International Group for the Psychology of Mathematics Education.* 2, 128–35.

Booth, L. (1984) *Algebra: Children's Strategies and Errors.* Windsor: NFER–Nelson.

Borovcnik, M., Bentz, H-J. and Kapadian, R. (1991) A probabilistic perspective. In Kapadia, R. and Brovcnik, M. (eds) *Chance Encounters: Probability in Education.* Dordrecht, The Netherlands: Kluwer Academic Publishers. 27–71.

Boulton-Lewis, G. M. (1987) Recent cognitive theories applied to sequential length measuring knowledge in young children. *British Journal of Educational Psychology.* 57, 330–42.

Boulton-Lewis, G. M., Cooper, T. J., Atweh, B., Pillay, H., Wilss, H. and Mutch, S. (1997) The transition from arithmetic to algebra: a cognitive perspective. *Proceedings of the 21st Conference of the International Group for the Psychology of Mathematics Education.* 2, 185–92.

Boulton-Lewis, G. M., Wilss, L. and Mutch, S. (1994) An analysis of young children's strategies and use of devices for length measurement. *Proceedings of the 18th Conference of the International Group for the Psychology of Mathematics Education.* 2, 128–35.

Boulton-Lewis, G. M., Cooper, T., Atweh, B., Pillay, H. and Wilss, L. (1998) Pre-algebra: a cognitive approach. *Proceedings of the 22nd Conference of the International Group for the Psychology of Mathematics Education.* 2, 144–51.

Bramald, R, (1994) Teaching probability. *Teaching Statistics.* 16, 3, 85–9.

Brenner, M. E. (1998) Meaning and money. *Educational Studies in Mathematics.* 36, 123–55.

Brito-Lima, A. P. and Da Rocha Falcão, J. T. (1997) Early development of algebraic representation among 6–13-year-old children: the importance of didactic contract. *Proceedings of the 21st Conference of the International Group for Psychology of Mathematics Education.* 2, 201–8.

Brousseau, G. (1988) Le contrat didactique: le milieu. *Recherches en Didactiques des Mathématiques.* 7.2, 33–115.

Brown, J. S., Collins, A. and Duguid, P. (1989) Situated cognition and the culture of learning. *Educational Researcher.* 18.1, 32–42.

Brown, M., Blondel, E., Simon, S. and Black, P. (1995) Dimensions of progression in measurement. *Proceedings of the Third British Congress of Mathematics Education (Part Two) Manchester Business School.* 160–7.

Brown, S. I. (1997) Thinking like a mathematician. *For the Learning of Mathematics.* 17, 2. 36–7.

Brown, S. I., Cooney, T. J. and Jones, D. (1990) Mathematics teacher education. In Houston, W. R. (ed.) *Handbook of Research on Teacher Education.* New York: Macmillan. 639–56.

Bryant, P. (1997) Mathematical understanding in the nursery school years. In Nunes, T. and Bryant, P. (eds) *Learning and Teaching Mathematics: An International Perspective.* 53–68.

Burger, W. and Shaugnessy, J. M. (1986) Characterizing the van Hiele levels of development in geometry. *Journal for Research in Mathematics Education.* 17, 31–4.

Burton, L. (1990) *Gender and Mathematics: An International Perspective.* London: Cassell.

Burton, L. (ed.) (1986) *Girls into Maths Can Go.* London: Holt, Rinehart and Winston.

Carpenter, T. P. and Moser, J. M. (1984) The acquisition of addition and subtraction concepts in Grades one through three. *Journal for Research in Mathematics Education.* 13.3, 179–202.

Carpenter, T. P. (1986) Conceptual knowledge as a foundation for procedural knowledge: implications from research on the initial learning of arithmetic. In Hiebert, J. (ed.) *Conceptual Knowledge and Procedural Knowledge: The Case of Mathematics.* Hillsdale, NJ: Lawrence Erlbaum Associates. 113–31.

Carpenter, T. P., Fennema, E. and Franke, M. L. (1993) *Cognitive Guided Instruction: Multiplication and Division.* Wisconsin Center for Educational Research, University of Wisconsin-Madison.

Carpenter, T. P., Fenneman, E., Peterson, P. L., Chiang, C. and Loef, M. (1989) Using knowledge of children's mathematical thinking in classroom teaching: an experimental study. *American Educational Research Journal.* 26, 449–531.

Case, R. (1992) *The Mind's Staircase.* Hillsdale, NJ: Erlbaum.

Cedillo Avalos, E. T. (1997) Algebra as language in use: a study with 11–12-year-olds using graphic calculators. *Proceedings of the 21st Conference of the International Group for the Psychology of Mathematics Education.* 2, 137–44.

Chaiklin, S. and Lesgold, S. (1984 April) *Prealgebra Students' Knowledge of Algebraic Tasks with Arithmetic Expressions.* Paper presented at the annual meeting of the American Educational Research Association, New Orleans, LA.

Chaiklin, S. (1989) Cognitive studies of algebra problem-solving and learning. In Wagner, S. and Kieran, C. (eds) *Research Issues in the Learning and Teaching of Algebra.* Reston, VA.: National Council of Teachers of Mathematics; Hillsdale, NJ: Lawrence Erlbaum. 93–114.

Chazan, D. and Bethell, S. C. (1994) Sketching graphs of an independent and dependent quantity: difficulties in learning to make stylized, conventional 'Pictures'. *Proceedings of the 18th Conference of the International Group for the Psychology of Mathematics Education.* II, 176–84.

Chevallard, Y. (1988) Sur l'analyse didactique. Deux études sur les notions de contrat et de situation. *Publications de l'IREM d'Aix-Marseille.* 14.

Chinnapan, M. (1998a) Schemes and mental models in geometry problem-solving. *Educational Studies in Mathematics.* 36, 201–17.

Chinnapan, M. (1998b) Restructuring conceptual and procedural knowledge for problem representation. *Proceedings of the 22nd Conference of the International Group for Mathematics Education 1998.* 2, 184–91.

Clarkson, P. C. and Dawe, L. (1997) NESB migrant students studying mathematics: Vietnamese students in Melbourne and Sydney. *Proceedings of the 21st Conference of the International Group for the Psychology of Mathematics Education.* 2, 153–60.

Clements, D. D., Sarama, J. and Swaminathan, S. (1997) Young children's concepts of shape. *Proceedings of the 21st Conference of the International Group for Mathematics Education.* 2, 161–8.

Clements, D. H. and Battista, M. T. (1989) Learning of geometric concepts in a Logo environment. *Journal for Research in Mathematics Education.* 20, 450–67.

Clements, D. H. and Battista, M. T. (1990) The effects of Logo on children's conceptualizations of angle and polygons. *Journal for Research in Mathematics Education.* 21, 356–71.

Clements, D. H. and Battista, M. T. (1992) Geometry and Spatial Reasoning. In Grouws, D. A. (ed.) *Handbook of Research on Mathematics Teaching and Learning.* New York: Macmillan Publishing Company. 420–64.

Clements, D. H., Battista, M. T., Sarama, J. and Swaminathan, S. (1996) Development of turn and turn measurement concepts in a computer-based Instructional Unit. *Educational Studies in Mathematics.* 31, 313–37.

Cobb, P. (1987a) Information-processing psychology and mathematics education – a constructivist perspective. *The Journal of Mathematical Behaviour.* 6, 3–40.

Cobb, P. (1987b) An investigation of young children's academic arithmetic contexts. *Educational Studies in Mathematics.* 18, 109–24.

Cobb, P., Wood, T. and Yackel, E. (1991) Learning through problem-solving: a constructivist approach to second grade mathematics. In von Glaserfeld, E. (ed.) *Radical Constructivism in Mathematics Education.* Dodrecht, The Netherlands: Kluwer. 157–76.

Cobb, P., Yackel, E. and Wood, T. (1992) Interaction and learning in mathematics classroom situations. *Educational Studies in Mathematics.* 2, 99–122.

Collis, K. F. and Romberg, T. A. (1991) Assessment of mathematical importance: an analysis of open-ended test items. In Wittrock, M. C. (ed.) *Cognition and Instruction.* Hillsdale, NJ: Erlbaum. 82–130.

Comiti, C. and Moreira Baltar, P. (1997) Learning process for the concept of area of planar regions in 12–13-year-olds. *Proceedings of the 21st Conference of the International Group for the Psychology of Mathematics Education.* 3, 264–71.

Committee of Inquiry into the Teaching of Mathematics in Schools (1982) *Mathematics Counts.* London: HMSO.

Confrey, J. (1994) Six approaches to functions using multi-representational software. *Proceedings of the 18th Conference of the International Group for the Psychology of Mathematics Education.* II, 217–24.

Cooney, T. J. (1994) Research and teacher education: in search of common ground.

Journal for Research in Mathematics Education. 25, 608–36.

Cooney, T. J. and Shealey, B. E. (1997) On understanding the structure of teachers' beliefs and the relationship to change. In Fennema, E. and Nelson, B. S. (eds) *Mathematics Teachers in Transition.* Hillsdale, NJ: Lawrence Erlbaum Associates. 87–109.

Cooper, T., Baturo, A. R. and Dole, S. (1998) Abstract schema versus computational proficiency in per cent problem-solving. *Proceedings of 22nd Conference of the International Group for the Psychology of Mathematics Education.* 2, 200–7.

Cortes, A. (1998) Implicit cognitive work in putting word problems into equation form. *Proceedings of the 22nd Conference of the International Group for the Psychology of Mathematics Education.* 2, 208–15.

Crawford, K., Gordon, S., Nicholas, J. and Prosser, M. (1994) Students' reports of their learning about functions. *Proceedings of the 18th Conference of the International Group for the Psychology of Mathematics Education.* II, 233–9.

Crowley, L., Thomas, M. and Tall, D. (1994) Algebra, symbols, and translation of meaning. *Proceedings of the 18th Conference of the International Group for the Psychology of Mathematics Education.* II, 240–7.

D'Ambrosio, U. (1985) Ethnomathematics and its place in the history and pedagogy of mathematics. *For the Learning of Mathematics.* 5.1, 44–8.

D'Ambrosio, B. S. and Newborn, D. S. (1994) Children's constructions of fractions and their implications for classroom instruction. *Journal of Research in Childhood Education.* 8, 150–61.

Davis, R. (1995) Why teach mathematics? Mathematics education and enactivist theory. *For the Learning of Mathematics.* 15.2, 2–9.

Davis, R. B. (1986) Conceptual and procedural knowledge in mathematics: a summary analysis. In Hiebert, J. (ed.) *Conceptual and Procedural Knowledge: The Case of Mathematics.* Hillsdale, NJ: Lawrence Erlbaum Associates. 265–300.

Davis, R. B. (1988) The interplay of algebra, geometry and logic. *Journal of Mathematical Behaviour.* 7, 9–28.

Davydov, V. V. (1962) An experiment in introducing elements of algebra in elementary school. *Soviet Education.* V (1), 27–37.

Davydov, V. V. (1988) Problems of developmental teaching. *Soviet Education.* 30, 6–97.

Davydov, V. V. (1990) *Soviet Studies in Mathematics Education: Volume 2. Types of Generalization in Instruction* (Kilpatrick, J. (ed.), trans. Teller, J.). Reston, VA: National Council for Teachers of Mathematics.

de Lange, J. (1996) Real problems with real world mathematics. *Proceedings of the 8th International Congress on Mathematical Education.* 83–110.

Department for Education and Employment (1999) *The National Numeracy Strategy.* Suffolk: Cambridge University Press.

Desforges, C. and Cockburn, A. (1987) *Understanding the Mathematics Teacher.* Lewes: Falmer Press.

de Villiers, M. (1998) To teach definitions in geometry or teach to define? *Proceedings of the 22nd Conference of the International Group for Mathematics Education.* 2, 248–55.

Dickson, L., Brown, M. and Gibson, O. (1984) *Children Learning Mathematics.* London: Cassell Educational Limited.

di Sessa, A. A. (1983) Phenomenology and the evolution of intuition. In Genter, D. and Stevens, A. (eds) *Mental Models.* Hillsdale, NJ: Lawrence Erlbaum Associates.

Doerr, H. M. (1998) Student thinking about models of growth and decay. *Proceedings of the 22nd Conference of the International Group for the Psychology of Mathematics Education.* 2, 256–63.

Doig, B. (1997) What makes scientific dialogue possible in the classroom? Paper presented at the *Multiple Perspectives for Scientific Dialogue: Implications for Classroom Practice* symposium at the 1997 Annual Meeting of the American Educational Research Association.

Dole, S., Cooper, T. C., Baturo, A. R. and Conoplia, Z. (1997) Year 8, 9 and 10 students' understanding and access of percent knowledge. *Mathematics Education Research Group of Australasia Incorporated.* 20, 147–54.

Dolk, M. and Uittenboogaard, W. (1989) De ouderavord. *Willem Bartjens.* 9, 1, 14–20.

Donaldson, M. (1978) *Children's Minds.* New York: W. W. Norton.

Douady, R. (1997) Didactic engineering. In Nunes, T. and Bryant, P. (eds) *Learning and Teaching Mathematics: An International Perspective.* Hove, East Sussex: Psychology Press Ltd.

Dubinsky, E. (1991) Reflective abstraction in advanced mathematical thinking. In Tall, D. (ed.) (1991) *Advanced Mathematical Thinking.* Dordrecht: Kluwer Academic Publishers.

Dunkels, A. (1992) Interweaving numbers, shapes, statistics and the real world in primary school and primary teacher education. *In* Robitaille, D. E. and Wheeler, D. H. (eds) *Selected Lectures from the 7th International Congress on Mathematical Education,* Les presses de l'Université de Laval, Québec. 123–37.

Eisenhart, M. A. (1988). The ethnographic research tradition and mathematics education research. *Journal for Research in Mathematics Education.* 19.2, 99–114.

Elwood, J. and Gipps, C. (1998) *Review of Recent Research on the Achievement of Girls in Single Sex Schools.* Institute of Education, University of London.

Empson, S. (1995) Using sharing situations to help children learn fractions. *Teaching Children Mathematics.* 2, 110–14.

Ernest, P. (1991) *The Philosophy of Mathematics Education.* Basingstoke, Hampshire: The Falmer Press.

Falk, R. (1993) Inference under uncertainty in conditional probabilities. In *Studies in Mathematics Education, Vol. 7. Teaching Statistics in Schools.* Paris: UNESCO.

Falk, R. and Konold, C. (1992) The psychology of learning probability. In Gordon, F. S. and Gordon, S. P. (eds) *Statistics for the Twenty-first Century.* The Mathematical Association of America.

Fasheh, M. (1982) Mathematics, culture and authority. *For the Learning of Mathematics.* 3.2, 2–8.

Feiman-Nemser, S. and Floden, R. E. (1986). The cultures of teaching. In Wottrock, M. C. (ed.) *Handbook of Research in Teaching* (3rd edition). London: Collier-Macmillan. 505–26.

Fennema, E. and Tartre, L. A. (1985) The use of spatial visualization in mathematics by girls and boys. *Journal for Research in Mathematics Education.* 16, 184–206.

Fennema, E. and Franke, M. L. (1992) Teachers' knowledge and its impact. In Grouws, D. A. (ed.) *Handbook of Research on Mathematics Teaching and Learning.* New York: Macmillan. 147–64.

Fischbein, E. (1987) *Intuition in Science and Mathematics.* Dordrecht, The Netherlands: Reidel.

Fischbein, E. (1994) The interaction between the formal, the algorithmic and the intuitive components in mathematical activity. In Biehler, R. *et al.* (eds) *Didactic of Mathematics as a Scientific Discipline.* Dordrecht: Kluwer Publishing Co.

Fischbein, E. and Gazit, A. (1984). Does the teaching of probability improve probabilistic intuitions? An exploratory research study. *Educational Studies in Mathematics.* 15, 1–24.

Fischbein, E., Nello, M. S. and Marino, M. S. (1991) Estimating odds and the concept of probability. *Educational Studies in Mathematics.* 22, 523–49.

Foster, R. (1994) Counting on success in simple addition tasks. *Proceedings of the 18th Conference of the International Group for the Psychology of Mathematics Education.* 2, 360–7.

Freudenthal, H. (1973) *Mathematics as an Educational Task.* Dordrecht, The Netherlands: Reidel.

Freudenthal, H. (1974) Soviet research on teaching algebra at the lower grades of the elementary school. *Educational Studies in Mathematics.* 5, 391–412.

Freudenthal, H. (1991) *Revisiting Mathematics Education.* Dordrecht: Kluwer Academic Publishers.

Furinghetti, F. and Paola, D. (1994) Parameters, unknowns and variables: a little difference? *Proceedings of the 18th Conference of the International Group for the Psychology of Mathematics Education.* II, 368–75.

Fuson, K. and Kwon, Y. (1991) Chinese-based regular and European irregular sytems of number words: the disadvantages for English-speaking children. In Durkin, K. and Shire, B. (eds) *Language in Mathematical Education.* Milton Keynes: Open University Press.

Fuson, K.C. (1992) Research on whole number addition and subtraction. In Grouws, D. A. (ed.) *Handbook of Research on Mathematics Teaching and Learning.* New York: Macmillan. 243–75.

Fuys, D., Geddes, D. and Tischler, R. (1988) The van Hiele model of thinking in geometry among adolescents. *Journal for Research in Mathematics Education Monograph.* 3.

Gal, H. and Vinner, S. (1997) Perpendicular lines – what is the problem? *Proceedings of the 21st Conference of the International Group for the Psychology of Mathematics Education.* 2, 281–8.

Gardiner, J. and Hudson, B. (1998) The evolution of pupils' ideas of construction and proof using hand-held dynamic geometry technology. *Proceedings of the 22nd Conference of the International Group for Mathematics Education.* 2, 337–40.

Garofalo, J. and Lester, F. (1985) Metacognition, cognitive monitoring and mathematical performance. *Journal for Research in Mathematics Education.* 16, 163–76.

Gates, P. (1997) The importance of social structure in developing a critical social psychology of mathematics education. *Proceedings of the 21st Conference of the International Group for the Psychology of Mathematics Education.* 2, 305–12.

Geary, D. C. (1996) Sexual selection and sex differences in mathematical abilities. *Behavioral and Brain Sciences.* 19, 224–47.

Gelman, R. and Meck, E. (1986) The notion of principle: the case of counting. In Hiebert, J. (ed.) *Conceptual Knowledge and Procedural Knowledge: The Case of Mathematics.* Hillsdale, NJ: Lawrence Erlbaum Associates Inc. 29–58.

Ginsburg, H. P., Choy, Y. E., Lopez, L. S., Netley, R. and Chao-Yuan, C. (1997) Happy birthday to you: early mathematical thinking of Asian, South American, and US children. In Nunes, T. and Bryant, P. (eds) *Learning and Teaching Mathematics: An International Perspective.* Hove, East Sussex: Psychology Press Limited.

Glencross, M. J. (1998) Developing a statistics anxiety rating scale. *Proceedings of the 22nd Conference of the International Group for the Psychology of Mathematics Education.* 4, 256.

Gomes-Ferreira, V. G. (1998) Conceptions as articulated in different microworlds exploring functions. *Proceedings of the 22nd Conference of the International Group for the Psychology of Mathematics Education.* 3, 9–16.

Goodchild, S. (1995) Seven dimensions of learning – a tool for the analysis of mathematical activity in the classroom. *Proceedings of the 19th Conference of the International Group for the Psychology of Mathematics Education.* (Recife, Brazil). 3, 113–20.

Graham, A. and Thomas, M. (1997) Tapping into algebraic variables through the graphic calculator. *Proceedings of the 21st Conference of the International Group for the Psychology of Mathematics Education.* 3, 9–16.

Gravemeijer, K. (1994) *Developing Realistic Mathematics Education.* Utrecht: CD-b Press.

Gravemeijer, K. (1997) Mediating between concrete and abstract. In Nunes, T. and Bryant, P. (eds) *Learning and Teaching Mathematics: An International Perspective.* Hove, Sussex: Psychology Press. 315–45.

Graves, B. and Zack, V. (1997) Collaborative mathematical reasoning in an inquiry classroom. *Proceedings of the 21st Conference of the International Group for the Psychology of Mathematics Education.* 3, 17–24.

Gray, E. (1994) Spectrums of performance in two digit addition and subtraction. *Proceedings of the 18th Conference of the International Group for the Psychology of Mathematics Education.* 3, 25–32.

Gray, E. M. and Tall, D. O. (1993) Success and failure in mathematics: the flexible meaning of symbols as process and concept. *Mathematics Teaching.* 142, 6–10.

Gray, E. M. and Tall, D. O. (1994) Duality, ambiguity and flexibility: a proceptual view of simple arithmetic. *Journal for Research in Mathematics Education.* 26, 115–41.

Gray, E., Pitta, D. and Tall, D. (1997) The nature of the object as an integral component of numerical processes. *Proceedings of the 21st Conference of the International Group for the Psychology of Mathematics Education.* 1, 115–30.

Green, D. (1983) A survey of probability concepts in 3000 pupils aged 11–16. In Grey, D. R., Holmes, C., Barnett, V. and Constable, G. M. (eds) *Proceedings of the First International Conference on Teaching Statistics.* Sheffield: Teaching Statistics Trust.

Green, D. R. (1987) Children's understanding of randomness: report of a survey of 1,600 children aged 7–11 years. In Davidson, R. and Swift, J. (eds) *Proceedings of the Second International Conference on Teaching Statistics.* 9, 1, 8–14.

Greeno, J. G. (1980) Some examples of cognitive task analysis with instructional implications. In Snow, R. E., Federico, P. and Montague, W. E. (eds) *Aptitude, Learning and Instruction, Volume 2: Cognitive Process Analyses of Learning and Problem-solving.* Hillsdale, NJ: Lawrence Erlbaum Associates. 1–21.

Greeno, J. G. (1987) Instructional representations based on research about understanding. In Schoenfeld, A. H. (ed.) *Cognitive Science and Mathematics.* Hillsdale, NJ: Lawrence Erlbaum Associates. 61–88.

Greer, B. (1992) Multiplication and division as models of situations. In Grouws, D. A. (ed.) *Handbook of Research on Mathematics Teaching and Learning.* New York: Macmillan. 276–95.

Groves, S. and Doig, B. (1998) The nature of the role of discussion in mathematics: three elementary teachers' beliefs and practice. *Proceedings of the 22nd Conference of the International Group for the Psychology of Mathematics Education.* 3, 17–24.

Gutiérrez, A., Jaime, A. and Fortuny, J. M. (1991) An alternative paradigm to evaluate the van Hiele levels. *Journal for Research into Mathematics Education.* 22, 237–51.

Haastrup, K. and Lindenskov, L. (1994). In Burton, L. (ed.) *Who Counts? Assessing Mathematics in Europe.* Stoke-on-Trent: Trentham Books Ltd. 23–40.

Halford, G. S. (1993) *Children's Understanding: The Development of Mental Models.* Hillsdale, NJ: Erlbaum.

Harper, E. (1987) Ghosts of Diophantus. *Educational Studies in Mathematics.* 18, 75–90.

Harries, T. and Sutherland, R. (1995) Access to algebra – the case of Logo. *Proceedings of the Third British Congress of Mathematics Education.* Manchester Business School. 208–15.

Hart, K. (ed.) (1981) *Children's Understanding of Mathematics 11–16.* London: John Murray.

Hart, K. and Sinkinson, A. (1988) Forging the link between practical and formal mathematics. *Proceedings of the 12th Conference of the International Group for the Psychology of Mathematics Education.* 23, 242–73.

Hasegawa, J. (1997) Concept formation of triangles and quadrilaterals in second grade. *Educational Studies in Mathematics.* 32, 157–79.

Hazzan, O. and Godenberg, O. (1997) An expression of the idea of successive refinement in dynamic geometry environments. *Proceedings of the 21st Conference of the International Group for Mathematics Education.* 3, 49–56.

Hembree, R. (1992) Experiments and relational studies in problem-solving: a meta analysis. *Journal for Research in Mathematics Education.* 26, 115–41.

Herscovics, N. and Linchevski, L. (1994) A cognitive gap between arithmetic and algebra. *Educational Studies in Mathematics.* 59–78.

Hershkowitz, R. (1989) Visualization in geometry – two sides of the coin. *Focus on Learning Problems in Mathematics.* 11, 61–76.

Hershkowitz, R, Ben-Chaim, D., Hoyles, C., Lappan, G., Mitchelmore, M. and Vinner, S. (1990) Psychological aspects of learning geometry. In Nesher, P. and Kilpatrick, J. (eds) *Mathematics and Cognition: A Research Synthesis by the International Group for the Psychology of Mathematics Education.* 70–95.

Hewitt, D. and Brown, E. (1998) On teaching early number through language. *Proceedings of the 22nd Conference of the International Group for the Psychology of Mathematics Education.* 3, 41–8.

Hiebert, J., Carpenter, T. P., Fennema, E., Fuson, K., Human P, Olivier, A. and Wearne, D. (1996) Problem-solving as a basis for reform in curriculum and instruction: the case of mathematics. *Educational Researcher.* 25.4, 12–22.

Hillel, J. and Kieran, C. (1988) Schemas used by 12-year-olds in solving selected turtle geometry tasks. *Recherches en Didactique des Mathématiques.* 8.12, 61–103.

Hoyles, C. and Sutherland, R. (1986) *When 45 equals 60.* London: University of London Institute of Education, Microworlds Project.

Hoyles, C. and Noss, R. (1996) *Windows on Mathematical Meanings: Learning Cultures and Computers.* Dordrecht: Kluwer.

Hughes, M. (1986) *Children and Number: Difficulties in Learning Mathematics.* Oxford: Basil Blackwood Ltd.

Jaworski, B. (1991) Interpretations of a constructivist philosophy in mathematics teaching. Unpublished doctoral dissertation. Milton Keynes: Open University.

Jaworski, B. (1994) The social construction of classroom knowledge. *Proceedings of the 18th Conference of the International Group for the Psychology of Mathematics Education.* 3, 73–80.

Jaworski, B. and Potari, D. (1998) Characterising mathematics teaching using the teaching triad. *Proceedings of the 22nd Conference of the International Group for the Psychology of Mathematics Education.* 3, 17–24.

Johnson, E. S. and Meade, A. C. (1987) Developmental patterns of spatial ability: an early sex difference. *Child Development.* 58, 725–41.

Johnson, S. (1996) The contribution of large scale assessment programmes to research on gender differences. *Educational Research and Evaluation.* 2.1, 25–49.

Jones, K. (1997) Children learning to specify geometrical relationships using a dynamic geometry package. *Proceedings of the 21st Conference of the International Group for Mathematics Education.* 3, 121–8.

Joseph, G. G. (1990) The politics of antiracist mathematics. *Proceedings of the First International Conference of the Group for the Political Dimensions of Mathematics Education.* University of London Institute of Education. 134–42.

Joseph, G. G. (1992) Different ways of knowing: contrasting styles of argument in India and the West. Robitaille, D. E. and Wheeler, D. H. (eds) *Selected Lectures from the 7th International Congress on Mathematical Education,* Les presses de l'Université de Laval, Québec. 183–98.

Julie, C. (1998) The production of artefacts as goal for school mathematics? *Proceedings of the 22nd Conference of the International Group for the Psychology of Mathematics Education.* 1, 49–65.

Kahneman, D. and Tversky, A. (1972) On prediction and judgment. *Oregon Institute Bulletin.* 12, 4.

Kaput J. J. (1987) Towards a theory of symbol use in mathematics. In Janvier, C. (ed.) *Problems of Representation in the Teaching and Learning of Mathematics.* London: Lawrence Erlbaum Associates, 159–96

Kaput, J. J. (1992) Technology and mathematics education. In Grouws, D. A. (ed.) *Handbook of Research on Mathematics Teaching and Learning.* New York: Macmillan Publishing Company. 515–56.

Kaput, J. (1993) The urgent need for proleptic research in the repesentation of quantitative relationships. In Romberg, T., Fennema, E. and Carpenter, T. (eds) *Integrating the Graphical Representations of Functions.* Hillsdale, NJ: Lawrence Erlbaum. 279–312.

Kaput, J. and Roschelle, J. (1997) Deepening the impact of technology beyond assistance with traditional formalisms in order to democratize access to ideas underlying calculus. *Proceedings of the 21st Conference of the International Group for the Psychology of Mathematics Education.* 1, 1105–13.

Kaur, B. (1990) Girls and mathematics in Singapore: the case of GCE 'O' Level mathematics. In Burton, L. (ed.) *Gender and Mathematics.* London: Cassell Education Ltd.. 98–112.

Kidman, G. and Cooper, T. J. (1997) Area integration rules for Grades 4, 6 and 8 students. *Proceedings of the 21st Conference of the International Group for the Psychology of Mathematics Education.* 132–43.

Kieran, C. (1986) Logo and the notion of angle among fourth and sixth grade children. In *Proceedings of PME 10.* London. 99–104.

Kieran, C. (1989) The early language of algebra: a structural perspective. In Wagner, S. and Kieran, C. (eds) *Research Issues in the Learning and Teaching of Algebra.* Virginia: National Council of Mathematics Teachers (Lawrence Erlbaum Associates). 159–96.

Kieran, C. (1992) The learning and teaching of school algebra. In Grouws, D. A. (ed.) *Handbook of Research on Mathematics Teaching and Learning.* New York: Macmillan. 390–419.

Kieran, C. (1994) A functional approach to the teaching of algebra: some pros and cons. *Proceedings of the 18th Conference of the International Group for the Psychology of Mathematics Education.* I, 157–75.

Kieran, C., Hillel, J. and Erlwanger, S. (1986) Perceptual and analytical schemas in solving structure turtle geometry tasks. In Hoyles, C., Noss, R. and Sutherland, R. (eds) *Proceedings of the Second Logo and Mathematics Educators Conference.* London: University of London. 154–61.

Kieren, T. (1988) Personal knowledge of rational numbers: its intuitive and formal development. In Hiebert, J. and Behr, M. (eds) *Number Concepts and Operations in the Middle Grades*. Reston, VA: National Council for Teachers of Mathematics. 53–92.

Kimura, D. (1992) Sex differences in the brain. *Scientific American*. September 1992. 81–7.

Kline, M. (1967) The theory of probability. In Kline, M. (ed.) *Mathematics for the Nonmathematician*. New York: Dover Publications. 520–40.

Knijnik, G. (1995) Intellectuals and social movements: examining power relations. *Numeracy, Race, Gender and Class: Proceedings of the Third International Conference of the Group for the Political Dimensions of Mathematics Education*. Bergen: Caspar Forlag A/S. 90–113.

Koehler, M. and Grouws, D. A. (1992) Mathematics teaching practices and their effects. In Grouws, D. A. (ed.) *Handbook of Research on Mathematics Teaching and Learning*. New York: Macmillan. 115–26.

Koirala, H. P. (1998) Preservice teachers' conceptions of probability in relation to its history. *Proceedings of the 22nd Conference of the International Group for the Psychology of Mathematics Education*. 3, 135–42.

Konold, C. (1991) Understanding students' beliefs about probability. In von Glasersfeld, E. (ed.) *Radical Constructivism in Mathematics Education*. Holland: Kluwer. 139–56.

Konold, C., Pollatsek, A., Well, A. and Gagnon, A. (1997) Students' analysis of data: research of critical barriers. In Garfield, J. B. and Burrill, G. (eds) *The Impact of New Technololgies in Learning and Teaching Statistics*. IASE Round Table Conference. Voorburg: International Association for Statistical Education. 151–67.

Kota, S. and Thomas, M. (1997) The promotion of algebraic problem-solving performance by affective factors. *Proceedings of the 22nd Conference of the International Group for the Psychology of Mathematics Education*. 3, 4–269.

Kuchemann, D. (1981) Algebra. In Hart, K. (ed.) *Children's Understanding of Mathematics: 11–16*. London: John Murray. 102–19.

Laborde, C. and Laborde, J-M. (1995) What about a learning environment where Euclidean concepts are manipulated with a mouse? In di Sessa, A. A., Hoyles, C. and Noss, R. (eds) *Computers and Exploratory Learning*. Berlin: Springer-Verlag. 48–67.

Lamon, S. J. (1998) Algebra: meaning through modelling. *Proceedings of the 22nd Conference of the International Group for the Psychology of Mathematics Education*. 3, 167–74.

Lampert, M. (1989) Choosing and using mathematical tools in classroom discourse. In Brophy, J. (ed.) *Advances in Research on Teaching*, 1. Greenwich: JAI Press Inc., CT. 233–64.

Lampert, M. (1990) When the problem is not the question and the solution is not the answer: mathematical knowing and teaching. *American Educational Research Journal*. 27, 29–63.

Laridon, P. E. and Glencross, M. J. (1994) *Proceedings of the 18th Conference of the International Group for the Psychology of Mathematics Education*. I, 50.

Lave, J. (1988) *Cognition in Practice: Mind, Mathematics and Culture in Everyday Life*. Cambridge: Cambridge University Press.

Lave, J. (1993) Situated learning in communities in practice. In Resnick, Lewin and Telsey (eds) *Perspectives in Shared Cognition*. American Psychology Association.

Lave, J. (1996) Teaching, as learning, in practice. *Mind, Culture and Activity*. 3.3, 149–64.

Lave, J. and Wenger, E. (1991) *Situated Learning: Legitimate Peripheral Participation*. New York: Cambridge University Press.

Lawrie, C. and Pegg, J. (1997) Some issues in using Mayberry's test to identify van Hiele levels. *Proceedings of the 21st Conference of the International Group for Mathematics Education.* 3. 184–91.

Lean, G. and Clements, M. A. (1981) Spatial ability, visual imagery and mathematical performance. *Educational Studies in Mathematics.* 12, 267–99.

Leder, G. (1992) Mathematics and gender: changing perspectives. In Grouws, D. A. (ed.) *Handbook of Research on Mathematics Teaching and Learning.* New York: Macmillan Publishing Company. 420–64.

Lee, L. and Wheeler, D. (1989) The arithmetic connection. *Educational Studies in Mathematics.* 20, 41–54.

Lemke, J. (1995) *Textual Politics: Discourse and Social Dynamics.* London: Taylor and Francis.

Lemut, E. and Greco, S. (1998) Re-starting algebra in high school: the role of systemic thinking and of representation systems command. *Proceedings of the 22nd Conference of the International Group for the Psychology of Mathematics Education.* 3, 91–8.

Lerman, S. (1990) Alternative perspectives of the nature of mathematics and their influence on the teaching of mathematics. *British Educational Research Journal* 16, 15–61.

Lerman, S. (1993) The position of the individual in radical constructivism: in search of the subject. In Malone, J. and Taylor, P. (eds) *Constructivist Interpretations of Teaching and Learning Mathematics.* Perth: Curtin University of Technology.

Lerman, S. (1998) A moment in the zoom of a lens: towards a discursive psychology of mathematics teaching and learning. *Proceedings of the 22nd Conference of the International Group for the Psychology of Mathematics Education.* 1, 66–84.

Lesh, R. (1987) The evolution of problem representations in the presence of powerful conceptual amplifiers. In Janvier, C. (ed.) *Problems of Representation in the Teaching and Learning of Mathematics.* London: Lawrence Erlbaum Associates. 197–206.

Lesh, R., Post, T. and Behr, M. (1987) Dienes revisited: Multiple embodiments in computer environments. In Wirzup, I. and Streit, R. (eds) *Development in School Mathematics Education Around the World.* Reston, VA: National Council of Teachers of Mathematics. 647–80.

Lester, F. K. (1988) Reflections about mathematical problem-solving research. In Charles, R. I. and Silver, E. A. (eds) *The Teaching and Assessing of Mathematical Problem-solving.* Reston, VA: National Council of Teachers of Mathematics. 115–24.

Linchevski, L. and Herscovics, N. (1994) Cognitive obstacles in pre-algebra. *Proceedings of the 18th Conference of the International Group for the Psychology of Mathematics Education.* III, 176–83.

Linchevski, L. and Herscovics, N. (1996) Crossing the cognitive gap between arithmetic and algebra: Operating on the unknown in the context of equations. *Educational Studies in Mathematics.* 30, 39–65.

Linchevski, L. and Kutscher, B. (1998) Tell me with whom you're learning, and I'll tell you how much you've learned: mixed-ability versus same-ability grouping in mathematics. *Journal for Research in Mathematics Education.* 29.5, 533–54.

Lins, B. (1998) Cabri as a cognitive tool. *Proceedings of the Day Conference of the British Society for the Learning of Mathematics held at King's College, London, Saturday 28th February 1998 and at the University of Birmingham.* 57–60.

Lins, R. C. (1994) Eliciting the meanings for algebra produced by students: knowledge, justification and semantic fields. *Proceedings of the 18th Conference of the*

International Group for the Psychology of Mathematics Education. II, 184–91.

MacGregor, M. and Stacey, K. (1996) Learning to formulate equations for problems. *Proceedings of the 20th Conference of the International Group for the Psychology of Mathematics Education.* 3, 289–96.

Magina, S. and Hoyles, C. (1997) Children's understanding of turn and angle. In Nunes, T. and Bryant, P. (eds) *Learning and Teaching Mathematics: An International Perspective.* Hove, East Sussex: Psychology Press Ltd.

Maher, C. A. Martino, A. M. and Davis R. B. (1994) Children's different ways of thinking about fractions. *Proceedings of the 18th Conference of the International Group for the Psychology of Mathematics Education.* 3, 208–15.

Mansfield, H. and Scott, J. (1990) Young children solving spatial problems. *Proceedings of the 14th Conference of the International Group for the Psychology of Mathematics Education.* 2, 275–82.

Marshall, S. P. (1995) *Schemas in Problem-solving.* New York: Cambridge University Press.

Mason, J. (1987) Representing representing. In Janvier, C. (ed.) *Problems of Representation in the Teaching and Learning of Mathematics.* London: Lawrence Erlbaum Associates, 159–96, 207–14.

Mayberry, J. (1983) The van Hiele levels of geometric thought in undergraduate preservice teachers. *Journal for Research in Mathematics Education.* 14.1, 58–69.

Mboyiya, T. (1998) Measuring project impact in the classroom: an analysis of learners' solution strategies. *Proceedings of the 22nd Conference of the International Group for the Psychology of Mathematics Education.* 4, 279.

Mellin-Olsen, S. (1987) *The Politics of Mathematics Education.* Dordrecht, The Netherlands: Kluwer Academic Publishers.

Mitchelmore, M. and White, P. (1998) Recognition of angular similarities between similar physical situations. *Proceedings of the 22nd Conference of the International Group for the Psychology of Mathematics Education.* 3, 271–8.

Monaghan, J. (1995) Student learning in a computer algebra environment: where are we and where do we go from here? *Proceedings of the Third British Congress of Mathematics Education.* Manchester Business School. 263–70.

Murray, H., Olivier, A. and Human, P. (1994) Fifth graders' multi-digit multiplication and division strategies after five years' problem-centered learning. *Proceedings of the 18th Conference of the International Group for the Psychology of Mathematics Education.* 3, 399–406.

Nemirovsky, R. (1993) *Motion, Flow, and Contours: The Experience of Continuous Change.* Unpublished doctoral dissertation. Harvard University.

Nemirovsky, R., Kaput, J. and Roschelle, J. (1998) Enlarging mathematical activity from modelling phenomena to generating phenomena. *Proceedings of the 22nd Conference of the International Group for the Psychology of Mathematics Education.* 3, 287–94.

Nesher, P. (1989) Microworlds in mathematical education: a pedagogical realism. In Reiniek, L. B. (ed.) *Knowing, learning and Instruction.* Hillsadle, NJ: Lawrence Erlbaum.

Neuman, D. (1997) Immediate and sequential experiences of numbers. *Proceedings of the 21st Conference of the International Group for the Psychology of Mathematics Education.* 3, 288–95.

Newstead, K. (1996) Language and strategies in children's solution of division problems. *Proceedings of the Day Conference, British Society for Research into Learning Mathematics.* 24 February 1996. 7–12.

Newstead, K. and Murray, H. (1998) Young students' construction of fractions. *Proceedings of the 22nd Conference of the International Group for the Psychology of Mathematics Education.* 3, 295–302.

Nickson, M. (1988) What is multicultural mathematics? In Ernest, P. (ed.) *Mathematics Teaching: The State of the Art.* Lewes, Sussex: Falmer Press. 236–41.

Nickson. M. (1992) The culture of the mathematics classroom: an unknown quantity? In Grouws, D. A. (ed.) *Handbook of Research on Mathematics Teaching and Learning.* New York: Macmillan. 100–14.

Nickson, M. (1995) Mathematics education as an international marketplace. In Kjaergard, T., Kvamme, A. and Linden, N. (eds) *Numeracy, Race, Gender and Class: Proceedings of the Third International Conference of the Group for the Political Dimensions of Mathematics Education.* Bergen: Caspar Forlag A/S. 150–68.

Nickson, M. (1998) What is the difference between a pizza and a relay race? The role of context in assessing mathematics. *British Journal of Curriculum and Assessment.* 7, 3, 19–22.

Nickson, M. and Lerman, S. (eds.) (1992) *The Social Context of Mathematics Education: Theory and Practice.* London: South Bank Press.

Nickson, M. and Prestage, S. (1994) England and Wales. In Burton, L. (ed.) *Who Counts? Assessing Mathematics in Europe.* Stoke-on-Trent: Trentham Books Ltd. 41–66.

Nielsen, L., Patronis, T. and Skovsmose, O. (1999) *Connecting Corners: A Greek–Danish Project in Mathematics Education.* Copenhagen: Forlaget Systime A/S.

Noddings, N. (1993) Politicizing the mathematics classroom. In Restivo, S., Van Bendegem, J. P. and Fischer, R. (eds) *Math Worlds: Philosophical and Social Studies of Mathematics and Mathematics Education.* Albany: State University of New York. 150–61.

Noss, R. (1987) Children's learning of geometrical concepts through Logo. *Journal for Research in Mathematics Education.* 18, 343–62.

Noss, R. (1988) The computer as a cultural influence in mathematical learning. *Educational Studies in Mathematics.* 19, 251–68.

Noss, R. and Hoyles, C. (1996) *Windows on Mathematical Meanings: Learning Cultures and Computers.* Dordrecht: Kluwer.

Nunes, T., Light, P., Mason, J. and Allerton, M. (1994a) *Children's Understanding of the Concept of Area.* London Institute of Education. Research Report presented to the ESRC.

Nunes, T., Light, P., Mason, J. and Allerton, M. (1994b) The role of symbols in structuring reason: studies about the concept of area. *Proceedings of the 18th Conference of the International Group for the Psychology of Mathematics Education.* 3, 255–62.

Nunes, T., Schliemann, A. and Carraher, D. W. (1993) *Street Mathematics and School Mathematics.* Cambridge: Cambridge University Press.

Oliveira, I. (1993) Rational numbers: strategies and misconceptions in sixth grade students. *Proceedings of the 18th Conference of the International Group for the Psychology of Mathematics Education.* 3, 392–8.

Orton, A. (1992) *Learning Mathematics. Issues, Theory and Practice* (Second Edition). London: Cassell.

Orton, J. (1997) Pupils' perception of pattern in relation to shape. *Proceedings of the 21st Conference of the International Group for the Psychology of Mathematics Education.* 3, 304–11.

Owens, K. and Outhred, L. (1997) Early representations of tiling areas. *Proceedings of the 21st Conference of the International Group for the Psychology of Mathematics Education.* 3, 312–19.

Pallaascho, R., Allaire, R. and Mongeau, P. (1993) The development of spatial competencies through alternating analystic and synthetic activities. *For the Learning of Mathematics.* 13, 3, 8–15.

Papert, S. (1994) Microworlds: transforming education. In Evans, S. and Clark, P. (eds) *The Computer Culture.* White River Press.

Parker, M. and Leinhart, G. (1995) Percent: a privileged proportion. *Review of Educational Research.* 65, 4, 421–81.

Patrick, H. (1990) *Gender Differences and Public Examination Results.* Paper presented at British Educational Research Association Conference.

Peck, D. M and Jencks, S. M. (1988) Reality, arithmetic, algebra. *Journal of Mathematical Behaviour.* 7, 85–91.

Pence, B. (1994) Teachers' perceptions of algebra. *Proceedings of the 18th Conference of the International Group for the Psychology of Mathematics Education.* IV, 17–24.

Piaget, J. and Inhelder, B. (1975) *The Origin of the Idea of Chance in Children.* London: Routledge and Kegan Paul.

Pike, C. D. and Forrester M. A. (1996) The role of number sense in children's estimating ability. British Society for Research into Learning Mathematics. *Proceedings of the Day Conference,* 9 November 1996. 43–8.

Pimm, D. (1995) *Symbols and Meanings in School Mathematics.* London: Routledge.

Pimm, D. (1996) Modern times: the symbolic surfaces of language, mathematics and art. In Puig, L. and Gutiérrez, A. *Proceedings of the 21st Conference of the International Group for the Psychology of Mathematics Education.* 3, 35–50.

Pirie, S., Martin, L. and Kieran, T. (1994) Mathematical images for fractions: help or hindrance? *Proceedings of the 18th Conference of the International Group for the Psychology of Mathematics Education.* 3, 247–54.

Popper, K. (1972) *Objective Knowledge – An Evolutionary Approach.* Oxford: Oxford University Press. 88–95.

Pratt, D. (1994) Active graphing. *Proceedings of the 18th Conference of the International Group for the Psychology of Mathematics Education.* IV, 57–64.

Pratt, D. (1995) Young children's active and passive graphing. *Journal of Computer Assisted Learning.* 11, 157–69.

Pratt, D. and Noss, R. (1998) The co-ordination of meanings for randomness. *Proceedings of the 22nd Conference of the International Group for the Psychology of Mathematics Education.* 4, 17–24.

Prawat, R. (1989) Promoting access to knowledge, strategy and disposition in students. *Review of Educational Research.* 59, 1–42.

Proceedings of the Third International Conference of the Group for the Political Dimensions of Mathematics Education. Bergen: Caspar Forlag A/S. 150–68.

Putnam, R. T., Lampert, M. and Peterson, P. L. (1990) Alternative perspectives on knowing mathematics in elementary schools. In Cazden, C. B. (ed.) *Review of Research in Education.* Washington, DC: American Educational Research Association. 57–150.

Rade, L. (1983) Stochastics at the school level in the age of the computer. In Grey, D. R., Holmes, P., Barnett, V. and Constable, G. M. *Proceedings of the First International Conference on Teaching Statistics.* Sheffield: Teaching Statistics Trust. 19–33.

Rade, L. (1984) Topic area: teaching statistics. *Proceedings of the Fifth International Congress on Mathematics Education.* Stuttgart: Birkhauser. 300–5.

Radford, I. (1994) Moving through systems of mathematical knowledge: from algebra with a single unknown to algebra with two unknowns. *Proceedings of the 18th Conference of the International Group for the Psychology of Mathematics Education.* IV, 73–80.

Redden, T. (1994) Alternative pathways in the transition from arithmetic thinking to algebraic thinking. *Proceedings of the 18th Conference of the International Group for the Psychology of Mathematics Education.* IV, 89–96.

Reggiani, M. (1994) Generalizations as a basis for algebraic thinking: observations with 11–12-year-old pupils. *Proceedings of the 18th Conference of the International Group for the Psychology of Mathematics Education.* IV, 97–104.

Resnick, L. B. and Omanson, S. F. (1984) *Learning to Understand Arithmetic.* Pittsburgh: University of Pittsburgh Learning Research and Development Center.

Restivo, S. (1993) The social life of mathematics. In Restivo, S., Van Bendegem, J. P. and Fischer, R. (eds) *Math Worlds: Philosophical and Social Studies of Mathematics and Mathematics Education.* Albany: State University of New York. 247–78.

Rico, L. (1994) Problems with duplicated semantic structure. *Proceedings of the 18th Conference of the International Group for the Psychology of Mathematics Education.* 4, 121–8.

Rogers, L. (1992) Mathematics education and social epistemology: the cultural origins of mathematics and the dynamics of the classroom. In Nickson, M. and Lerman, S. (eds) *The Social Context of Mathematics Education: Theory and Practice.* London: South Bank Press. 149–59.

Rossouw, L. and Smith, E. (1997) Teachers' pedagogicial knowledge of geometry. *Proceedings of the 21st Conference of the International Group for Mathematics Education.* 3, 57–65.

Ruthven, K. (1998) The use of mental, written and calculator strategies of numerical computation by upper primary pupils within a 'calculator-aware' number curriculum. *British Educational Research Journal.* 24, 21–42.

Ruwisch, S. (1998) Children's multiplicative problem-solving strategies in real-world situations. *Proceedings of the 22nd Conference of the International Group for the Psychology of Mathematics Education.* 4, 73–80.

Saad, S. and Davis, G. (1997) Spatial abilities, van Hiele levels and language used in three dimensional geometry. *Proceedings of the 21st Conference of the International Group for Mathematics Education.* 4, 104–11.

Saenz-Ludlow, A. and Waldgrave, C. (1998) Third Graders' interpretations of equality and the equal symbol. *Educational Studies in Mathematics.* 35, 153–87.

Saxe, G. B. (1991) *Culture and Cognitive Development: Studies in Mathematical Understanding.* New Jersey: Lawrence Erlbaum Associates.

Schaeffer, R. (1992) Statistics in the school and college curriculum. In Gaulin, C., Hodgson, R., Wheeler, D. H. and Egsgard, J. C. (1992) *Proceedings of the 7th International Congress on Mathematical Education.* Québec: Les presses de l'Université de Laval. 286–8.

Schliemann, A., Araujo, C., Cassundé, M.A., Macedo, S. and Nicéas, L. (1994) School children versus street sellers' use of the commutative law for solving multiplication problems. *Proceedings of the 18th Conference of the International Group for the Psychology of Mathematics Education.* 4, 209–16.

Schliemann, A. and Carraher, D. (1992) Proportional reasoning in and out of school. In Light, P. and Butterworth, G. (eds) *Context and Cognition: Ways of Learning and Knowing.* Hillsdale, NJ: Lawrence Erlbaum Associates. 47–73.

Schoenfeld, A. H. (1986) On having and using geometric knowledge. In Hiebert, J. (ed.) *Conceptual and Procedural Knowledge: The Case of Mathematics.* Hillsdale, NJ: Lawrence Erlbaum Associates. 225–64.

Schoenfeld, A. H. (1987) What's all the fuss about metacognition? In Shoenfeld, A. (ed.) *Cognitive Science and Mathematics Education.* Hillsdale, NJ: Lawrence Erlbaum Associates. 189–216.

Schoenfeld, A. H. (1992) Learning to think mathematically: problem-solving, metacognition, and sense making in mathematics. In Grouws, D. A. (ed.) *Handbook of Research on Mathematics Teaching and Learning.* New York; Macmillan. 334–70.

Schoenfeld, A. (1996) On having and using geometric knowledge. In Hiebert, J. (ed) *Conceptual and Procedural Knowldge: The Case of Mathematics.* Hillsdale, NJ: Lawrence Erlbaum Associates. 225–64.

Schubauer-Leoni, M. L. and Perret-Clermont, A. N. (1997) Social interactions and mathematics learning. In Nunes, T. and Bryant, P. (eds) *Learning and Teaching Mathematics: An International Perspective.* Hove, East Sussex: Psychology Press Ltd. 265–83.

Scott-Hodgetts, R. (1992) The National Curriculum: implications for the sociology of the classroom. In Nickson, M. and Lerman, S. (eds) *The Social Context of Mathematics Education: Theory and Practice.* London: South Bank Press. 10–25.

Sfard, A. (1987) Two conceptions of mathematical notions: operational and structural. In Bergeron, J. C., Herscovics, N. and Kieran, C. (eds) *Proceedings of the 11th Conference of the International Group for the Psychology of Mathematics Education.* III, 162–9.

Sfard, A. (1989) Transition from operational to conceptual construction: the notion of function revisited. In Vergnaud, G., Rogalski, J. and Artigue, M. (eds) *Proceedings of the 13th Conference of the International Group for the Psychology of Mathematics Education.* III, l62–9.

Sfard, A. (1991) On the dual nature of mathematical conceptions: reflections on processes and objects as different sides of the same coin. *Educational Studies in Mathematics.* 22, 1–36.

Sfard A. and Linchchevski (1994) The gains and pitfalls of reification – the case of algebra. *Educational Studies in Mathematics.* 26, 191–228.

Shaugnessy, J. M. (1992) Research in probability and statistics: Reflections and directions. In Grouws, D. A. (ed.) *Handbook on Research in the Teaching and Learning of Mathematics.* New York: Macmillan. 465–94.

Sierpsinka, A. (1992) *Lecture Notes on Understanding in Mathematics.* Unpublished paper, Concordia University, Department of Mathematics, Montréal, Québec.

Skovsmose, O. (1994) *Towards a Philosophy of Critical Mathematics Education.* Dodrecht, The Netherlands: Kluwer Academic Publishers.

Soh, L. (1994) Pupils' understanding of beginning algebra. *Proceedings of the 18th Conference of the International Group for the Psychology of Mathematics Education.* I, 111.

Sowell, E. J. (1989) Effects of manipulative materials in mathematics instruction. *Journal for Research in Mathematics Education.* 20, 498–505.

Spradley, J. (1980) *Participant Observation.* New York: Holt, Rinehart and Winston.

Springer, S. P. and Deutsch, G. (1981) *Left Brain, Right Brain.* New York: W. H. Freeman and Company.

Stacey, K. and MacGregor, M. (1994) Algebraic sums and products: students' concepts and symbolism. *Proceedings of the 18th Conference of the International Group for the Psychology of Mathematics Education.* IV, 289–96.

Steinberg, R. (1994) Children's invented strategies and algorithms in division. *Proceedings of the 18th Conference of the International Group for the Psychology of Mathematics Education.* 4, 305–12.

Streefland, L. (1991a) *Fractions in Realistic Mathematics Education: A Paradigm of Developmental Research.* Dordrecht, The Netherlands: Kluwer Academic Publishers.

Streefland, L. (1991b) Fractions: An integrated perspective. In Streefland, L. (ed.) *Realistic Mathematics Education in Primary School.* The Netherlands: Freudenthal Institute. 93–118.

Streefland, L. (1994) Pre-algebra from a different perspective. *Proceedings of the 18th Conference of the International Group for the Psychology of Mathematics Education.* I, 74.

Streefland, L. (1997) Charming fractions or fractions being charmed? In Nunes, T. and Bryant, P. (eds) *Learning and Teaching Mathematics: An International Perspective.* Hove, Sussex: Psychology Press. 347–71.

Sutherland, R. (1987) A study of the use and understanding of algebra related concepts within a Logo environment. In Bergeron, J. C., Herscovics, N. and Kieran, C. (eds) *Proceedings of the 11th Conference of the International Group for the Psychology of Mathematics Education.* I, 241–7.

Tall, D. O. and Thomas, M. O. (1991) Encouraging versatile thinking in algebra using the computer. *Educational Studies in Mathematics.* 22, 191–228.

Thomas, M. O. and Tall, D. O. (1988) Longer term conceptual benefits from using a computer in algebra teaching. *Proceedings of the 12th Conference of the International Group for the Psychology of Mathematics Education.* Veszprem, Hungary.

Thorpe, J. (1989) Algebra: what should we be teaching and how should we teach it? In Wagner, S. and Kieran, C. (eds) *Research Issues in the Learning and Teaching of Algebra.* Virginia: National Council of Mathematics Teachers. Lawrence Erlbaum Associates.

Tomazos, D. (1997) Investigating change in a primary mathematics classroom: valuing the students' perspective. *Proceedings of the 21st Conference of the International Group for the Psychology of Mathematics Education.* 4, 222–9.

Tomlinson, S. (1987) Curriculum option choices in multi-ethnic schools. In Tryona, B. (ed.) *Racial Inequality in Education.* London: Tavistock. 92–108.

Tooley J. and Darby, D. (1999) *Educational Research: A Critique Report* presented to Office for Standards in Education.

Truran, J. M. (1998) Using research into children's understanding of the symmetry of dice in order to develop a model of how they perceive the concept of a random generator. *Proceedings of the 22nd Conference of the International Group for the Psychology of Mathematics Education.* 4, 121–8.

Truran, K. and Ritson, R. (1997) Perceptions of unfamiliar random generators – links between research and teaching. *Proceedings of the 21st Conference of the International Group for the Psychology of Mathematics Education.* 4, 238–45.

Tsamir, P., Almog, N. and Tirosh, D. (1998) Students' solutions of inequalities. *Proceedings of the 22nd Conference of the International Group for the Psychology of Mathematics Education.* 4, 129–36.

Underhill, R. G. (1991) Two layers of constructivist interaction. In von Glaserfeld, E. (ed.) (1991) *Radical Constructivism in Mathematics Education.* Dordrecht: Kluwer Academic Publishers. 229–48.

Usiskin, Z. (1982) *Van Hiele Levels and Achievement in Secondary School Geometry (Final report of the Cognitive Development and Achievement in Secondary School Project).* Chicago, Ill.: University of Chicago, Department of Education.

Usiskin, Z (1988) Conceptions of school algebra and uses of variables. In Coxford, A. F. and Shulbert, A. P. (eds) *The Ideas of Algebra K–12.* Reston, Virginia: NCTM 8–19.

van Hiele, P. M. (1986) *Structure and Insight.* Orlando: Academic Press.

Vergnaud, G. (1997) The nature of mathematical concepts. In Nunes, T. and Bryant, P.

(eds) *Learning and Teaching Mathematics: An International Perspective.* Hove, East Sussex: Psychology Press Ltd. 5–26.

von Glaserfeld, E. (1991)(ed.) *Radical Constructivism in Mathematics Education.* Dordrecht: Kluwer Academic Press.

Voss, J. F., Perkins, D. N. and Segal, J. W (1993) (eds) *Informal Reasoning and Education.* Hillsdale, NJ: Erlbaum.

Vygotsky, L. (1962) *Thought and Language.* Cambridge, Mass.: MIT Press.

Vygotsky, L.S. (1978) *Mind in Society.* Cambridge: Harvard University.

Watson, A. (1998) What makes a mathematical performance noteworthy in informal teacher assessment? *Proceedings of the 22nd Conference of the International Group for the Psychology of Mathematics Education.* 4, 169–76.

Watson, J. M. and Collis, K. F. (1994) Mutimodal functioning in understanding chance and data concepts. *Proceedings of the 18th Conference of the International Group for the Psychology of Mathematics Education.* IV, 369–76.

Wertsch, J. V. (1991) *Voices of the Mind: A Sociocultural Approach to Mediated Action.* London: Harvester.

Wheatley, G. and Reynolds, A. (1996) The construction of abstract units in geometric and numeric settings. *Educational Studies in Mathematics.* 30, 67–83.

Wigley, A. (1997) Approaching number through language. In Thompson, I. (ed.) *Teaching and Learning Early Number.* Buckingham: Open University Press. 113–22.

Wilensky, U. J. (1993) *Connected Mathematics: Building Concrete Relationships with Mathematical Knowledge.* Doctoral thesis. Massachusetts Institute of Technology.

Williams, E. M. and Shuard, H. (1976) *Primary Mathematics Today.* London: Longman Group Limited.

Wilson, M. (1990) Measuring a van Hiele geometry sequence: a reanalyssis. *Journal for Research into Mathematics Education.* 3, 230–41.

Winter, R. (1992) Mathophobia, Pythagoras and roller-skating. In Nickson, M. and Lerman, S. (eds) *The Social Context of Mathematics Education: Theory and Practice.* London: South Bank Press. 81–93.

Womack, D. and Williams, J. (1998) Intuitive counting strategies of 5–6-year-old children within a transformational arithmetic framework. *Proceedings of the 22nd Conference of the International Group for the Psychology of Mathematics Education.* 4, 193–200.

Wong, M. P. H. (1997) Numbers versus letters in algebraic manipulation: which is more difficult? *Proceedings of the 21st Conference of the International Group for the Psychology of Mathematics Education.* 4, 285–90.

Wright, R. J. (1998) Children's beginning knowledge of numerals and its relationship to their knowledge of number words: an exploratory, observational study. *Proceedings of the 22nd Conference of the International Group for the Psychology of Mathematics Education.* 4, 201–8.

Yerushalmy, M. (1997) Emergence of new schemes for solving algebra word problems: the impact of technology and the function approach. *Proceedings of the 21st Conference of the International Group for the Psychology of Mathematics Education.* 1, 165–78.

Zack, V. (1994) Vygotskyan applications in the elementary mathematics classroom: looking to one's peers for helpful explanations. *Proceedings of the 18th Conference of the International Group for the Psychology of Mathematics Education.* 4, 409–16.

Zack, V. (1997) 'You have to prove us wrong': proof at the elementary school level. *Proceedings of the 21st Conference of the International Group for the Psychology of Mathematics Education.* 4, 291–8.

Zaslavsky, C. (1989) Integrating mathematics with the study of cultural traditions. In Keitel, C., Damerow, P., Bishop, A. J. and Gerdes, P. (eds) *Mathematics Education and Society.* Paris: United Nations Educational, Scientific and Cultural Organization (UNESCO). 14–15.

Index